T0297962

ELEMENTS OF PETROLEUM SCIENCE

Hamid N. Alsadi

ISBN
978-1-5437-4814-7 (sc)
978-1-5437-4815-4 (e)

Library of Congress Control Number: 2018960160

Print information available on the last page.

To order additional copies of this book, contact
Toll Free 800 101 2657 (Singapore)
Toll Free 1 800 81 7340 (Malaysia)
www.partridgepublishing.com/singapore
orders.singapore@partridgepublishing.com

10/16/2018

PARTRIDGE

PREFACE

A multidisciplinary science, such as science of petroleum, requires wide diversity of specializations in order to be able to give a complete presentation. Petroleum science deal with pure sciences as Physics, Chemistry, and Geology in addition to applied sciences as engineering and computer technologies. In view of this status, full and complete coverage of the subject given in a limited-size book, becomes a real challenge.

The motive for accepting the challenge and writing this book stemmed from my long time (over 35 years) experience of working in the field of petroleum exploration. From my past technical services (in both of the academic and industrial worlds) I found that there is need for a concise and self-contained monograph giving a summary for the essential aspects of the science of petroleum. The subject matter should cover definition of the petroleum substance, its generation, migration from the source zone and accumulation into a subsurface oil reservoir. This is exactly what I did.

This book, which is mostly a compilation of information obtained from various scientific sources, is designed to be a teaching text serving interested audiences who are not necessarily equipped with high-level scientific specializations. It consists of eight chapters. The first chapter is devoted to definition of petroleum as a substance, followed by three chapters on Geology, oil generation, migration, and accumulation into oil reservoirs. The next three chapters present oil exploration techniques, oil-well drilling and production. Chapter-8 covers definitions for oil transport means and storage.

I wish to acknowledge the assistance received from my family (my wife Asira, sons: Majid, Muhannad, and daughter Thuraya) during my work in this project. Our family IT, my son Eng. Mahir, helped a lot in text editing and in drawing of most of the figures.

Hamid Nassar Alsadi
16. 6. 2018

Contents

Chapter 1

1. THE PETROLEUM SUBSTANCE

1.1 Basic Definitions

The term " petroleum" is derived from the medieval Latin language to mean rock-oil; *petra* is rock, and *oleum* is oil. Usually, either petroleum, or oil, is used to refer to the naturally occurring petroleum matter which is essentially made up of a mixture of hydrocarbon compounds. In addition to this general usage of these terms, there are special cases where the term "oil" is used for the naturally occurring liquid petroleum, namely crude oil, to differentiate it from the gaseous petroleum (natural gas) and solid petroleum (asphalt, pitch, or bitumen).

The oil matter is believed to have been generated alongside the precipitation process of a fine-grained material of rocks (of argillaceous or calcareous nature) containing organic matter that is later transformed into a more homogeneous organic matter of complex chemical composition called kerogen. This process is thought to take place with the help of a type of bacteria that acts under anaerobic environments and under favorable pressure and temperature conditions. At a later stage, the kerogen changes into hydrocarbon matter. The place where the transformation process occurs is usually referred to as the oil kitchen; the rock medium within which oil—the hydrocarbon matter—is formed is called the source rocks or mother rocks.

After the generation phase, and due to the effective overburden pressure, the generated oil migrates from the mother rocks to the porous neighboring rock media. This motion, called primary oil migration, is followed by a second migration process, where the oil continues in motion (under favorable conditions of porous and permeable pathways) to the final accumulation zones. The migrating oil is accumulated in a rock zone when its motion is stopped and prevented from further movement. The closed zone in which oil is trapped is normally referred to as the oil trap.

1.2 Chemistry of Petroleum

The chemical composition of the petroleum matter is simple. It consists of two basic elements: hydrogen (H) and carbon (C). Together they form a chemical compound called hydrocarbon. In its simplest form, thc hydrocarbon compound is represented by the molecule CH_4.

It is interesting to note the striking similarity between oil and water. Like oil, water consists of two chemical elements, namely hydrogen and oxygen forming the water molecule (H_2O). Further, both water and oil exist in nature in liquid state mostly and in gaseous or solid states occasionally. In nature, both substances can be found in rock pores and on the earth's surface. Water prevails on the earth's surface as oceans, seas, lakes, and rivers, and to a lesser extent as underground accumulations present in rock pores. Petroleum, on the other hand, exists normally as accumulations existing in rock pores and, in certain instances, as surface seepages and solid tar deposits. With fundamental economic roles in life, oil and water became active elements in political relationships which may lead, under certain circumstances, to military confrontations.

1.2.1 The Hydrocarbon Compound

Due to the fact that carbon compounds and water are present in all living organisms, the carbon compounds are usually called hydrocarbons. For the same reason, the branch of science concerned with the study of the carbon chemical properties is called the science of organic chemistry.

One of the important chemical characteristics of the carbon atoms is their affinity to form chain compounds of different types and sizes. Some chains are open (linear series), and others are closed (closed series or rings). These compounds are much more abundant than any other type of chemical compounds. In fact, the carbon element is found in about 90 per cent of all chemical compounds known at present. It is estimated that the number of compounds in which carbon enters their composition are more than two million. This huge number is continually increasing with time since there are about a hundred thousand new compounds devised every year (*Prentice Hall Science Book*, 1995, p. 232).

1.2.2 The Hydrocarbon Molecule

As explained earlier, the hydrocarbon molecule consists basically of the two elements carbon (C) and hydrogen (H). these two elements are chemically combined to form a molecule of the form (mC + nH), where m and n are integers. According to the numbers m and n and ways of atom bonding, the hydrocarbon molecule takes different forms and sizes. Since the valence of the carbon atom is equal to 4 and the hydrogen valence is 1, the simplest hydrocarbon compound is of the form (CH_4) which represents the hydrocarbon gas methane.

```
      H
      |
  H — C —H
      |
      H
```

The simplest hydrocarbon compound, Methane (CH_4), made up of one carbon and four hydrogen atoms.

1.2.3 The Hydrocarbon Series

As stated previously, the hydrocarbon molecule has chains of carbon and hydrogen atoms bonded together in accordance to organic chemistry laws and principles. In general, the hydrocarbon molecule may be of a series of atoms forming straight chains which can be open (simple or branched) or a close ring-type series. These chains, called hydrocarbon series, may be saturated compounds when all four bonds of the carbon atoms in the series are engaged with hydrogen atoms. In cases where some of the carbon atoms are not fully engaged with hydrogen atoms, the series is termed "unsaturated hydrocarbon."

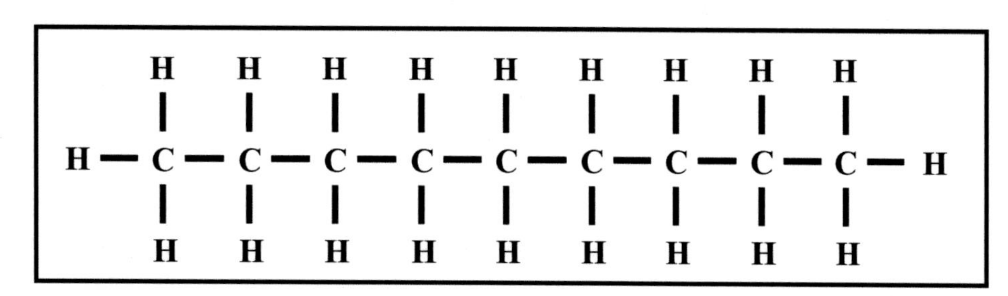

Example of a simple saturated hydrocarbon series ($C_9 H_{20}$), made up of nine carbon atoms and twenty hydrogen atoms.

An example of the unsaturated hydrocarbon compound is the ethylene (C_2H_4) which, when saturated, becomes the saturated compound ethane (C_2H_6). These two corresponding hydrocarbons are represented as follows.

(Ethane, C_2H_6) (Ethylene, C_2H_4)

```
    H  H              H  H
    |  |              |  |
H - C - C - H         C = C
    |  |              |  |
    H  H              H  H
```

Examples of: saturated (Ethane, C_2H_6) and unsaturated (Ethylene, C_2H_4), hydrocarbon series.

In addition to the saturation property, the hydrocarbon molecule has another important property. It is the branching property. For the same number of carbon atoms, a saturated hydrocarbon may be a simple and straight series or may be of a branched form. In order to differentiate between these two forms, the branched compound is called an isomer. In general, hydrocarbon compounds with the same molecular formula but with different structural form are called isomers. For example, the hydrocarbon butane (C_4H_{10}) and its corresponding isomer are presented as follows.

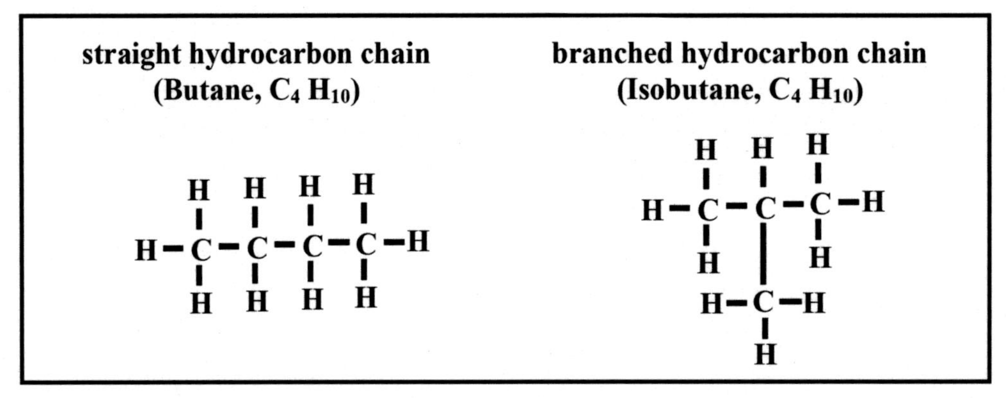

Examples of: straight hydrocarbon chain (Butane, C_4H_{10}) and the corresponding branched hydrocarbon chain; Butane isomer (or Isobutane).

The Ring-Form Hydrocarbon

Another type of the hydrocarbon molecule is ring form which can be of saturated or unsaturated nature. The simplest ring-type hydrocarbon is the saturated cyclohexane (C_6H_{12}) and its unsaturated cyclohexene (C_6H_6), commonly known by the name benzene. The two forms are presented as follows.

saturated ring-type hydrocarbon (C_6H_{12}) | **unsaturated ring-type hydrocarbon (C_6H_6)**

Examples of: the ring-type saturated hydrocarbon (Cyclohexane, C_6H_{12}) and the corresponding unsaturated hydrocarbon (C_6H_6)

1.2.4 Classification of Hydrocarbon Series

As mentioned earlier in the chapter, hydrocarbon compounds can be in the form of straight chains of carbon atoms or cyclic chains with or without branches. In all cases, the series may be saturated or unsaturated. Classification of the hydrocarbon compounds may be summarized by the following figure (fig. 1.1).

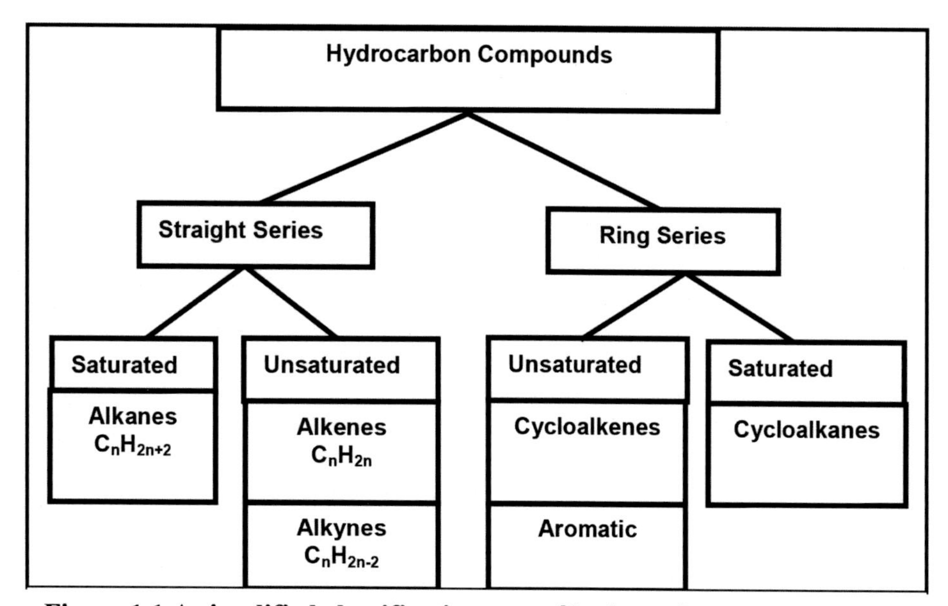

Figure 1.1 A simplified classification way of hydrocarbon compounds.

It is apparent from this figure that hydrocarbon compounds can be divided into two main groups: straight series and ring-form series. Each group is subdivided into saturated and unsaturated series. The saturated straight series are termed "alkanes," while the unsaturated straight series are termed "alkenes." The corresponding terms applied for the saturated and unsaturated ring series are "cycloalkanes" and "cycloalkenes," respectively.

Brief definitions of these divisions are presented here.

The Alkanes Series

This series (also called paraffin series) is an open-saturated chain which can be simple, straight, or branched chain of carbon atoms. Because of the saturation property, alkanes are generally not chemically active substances. For example, they are non-reactive with acids and alkalis. For this reason, these compounds are also called paraffins. This term is derived from the Latin words *parum* ("little") and *affinis* ("affinity"). The exception is that they are liable to burning (chemical reaction with oxygen) and liable to undergo atom substitution (atom replacement) as, for example, replacement of a hydrogen atom by a chlorine atom (Goddard and Hotton, 1955, p. 291).

The simplest alkane is the methane compound (CH_4) which consists of one carbon atom. Next to methane in complexity are the multi-carbon alkanes as C_2H_6, C_3H_8, C_4H_{10}, and so on. The general form of these compounds is C_nH_{2n+2}, where the integer $n \geq 1$ represents the number of carbon atoms in the molecule. Thus, for the *n* values 1, 2, 3, 4, the corresponding hydrocarbons will be the hydrocarbons: methane (CH_4), ethane (C_2H_6), propane (C_3H_8), and butane (C_4H_{10}).

In addition to the chemical properties of a hydrocarbon, the physical properties depend on the molecule structural form and its size, measured by the number of carbon and hydrogen atoms. For example, a paraffin hydrocarbon at standard pressure and temperature is at liquid state when the number of carbon atoms is in the range of five to fifteen. Paraffins in which the number of carbon atoms is fewer than five are found in a gaseous state. When the number exceeds twenty, the hydrocarbon becomes waxier in nature and solid for larger carbon numbers. Thus, the hydrocarbons methane, ethane, propane, and butane are gases; pentane, hexane, octane, and with a carbon number of up to fifteen atoms are liquid. Hydrocarbons having carbon numbers exceeding twenty, as in the compound $C_{20}H_{42}$, are solids (Chapman, 1976, p. 27). Dependence of the physical state of a paraffin hydrocarbon on number of carbon atoms is shown in Fig. 1.2.

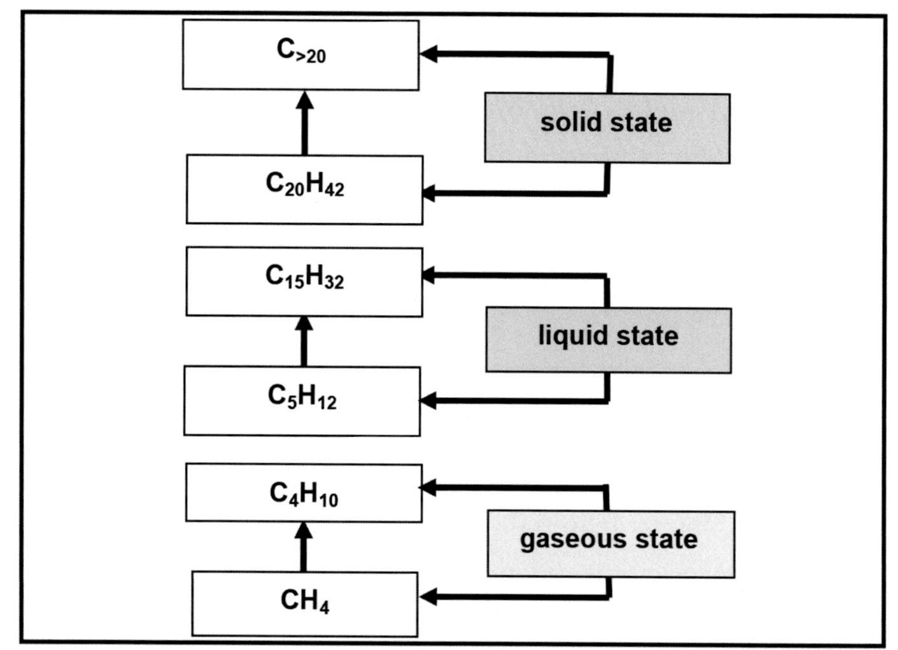

Figure 1.2 The physical state of a paraffin hydrocarbon depends on the number of carbon atoms present in the molecule.

Paraffin compounds form about 15 per cent to 20 per cent of normal crude oil. This ratio increases to 35 per cent in light oils and decreases to very low levels in heavy oils (Banks and King, 1986, p. 283).

The Alkenes Series

The alkenes hydrocarbons are unsaturated straight-chain compounds which have one double bond connecting carbon atoms. The chemical molecular formula is C_nH_{2n}. It is found in small proportions in crude oils. These compounds are also called olefins because they constitute oily liquids when they react with chlorine gases. Being unsaturated, the alkenes are highly active in chemical reactions. The simplest alkene hydrocarbon is ethylene (C_2H_4), whose structure is presented as follows.

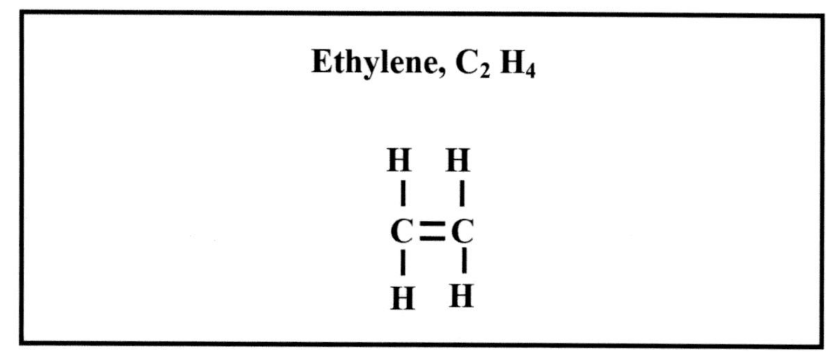

Example of the Alkene hydrocarbon the (Ethylene, C_2H_4) .

Under high-pressure conditions, alkene compounds can, by addition process (called polymerization), form larger molecules. For example, the ethylene can, by this method, form polyethylene (usually called polythene). With this type of reaction, the plastic substance (synthetic rubber) and the glass-like hard plastics (Perspex) are manufactured. Other such polymers commonly found in industry are polypropylene and polystyrene.

The Alkynes Series

Like alkenes, alkynes are unsaturated straight-chain compounds but with triple bonds connecting carbon atoms. The chemical molecular formula is C_nH_{2n-2}. These compounds are also called acetylene series because the first compound in this series is the compound acetylene, an important industrial material. Alkynes are highly active in chemical reactions. The simplest alkyne hydrocarbon is acetylene (C_2H_2), whose structure is presented as follows.

Acetylene, C_2H_2

$H-C\equiv C-H$

Example of the Alkyne hydrocarbon, the (Acetylene, C_2H_2).

The Cycloalkanes Series

This is an Alkane-analoge type of hydrocarbon in which the molecules take up the form of ring structures which are called Cycloalkane or Cycloparaffins as they are sometimes called. These are saturated compounds of molecular structural form (C_nH_{2n}) with single carbon bonds. The Cyclobutane and Cyclopentane are examples, shown as follows:

(Cyclobutane C_4H_8) **(Cyclopentane C_5H_{10})**

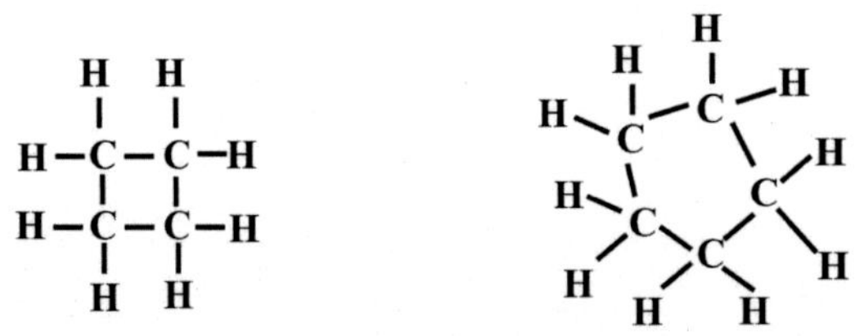

Examples of the saturated ring-form hydrocarbons, Cycloalkane series, (Cyclobutane C_4H_8) and (Cyclopentane C_5H_{10}).

In the industry the **Cycloalkanes** are referred to as **Naphthenes** which are darker in colour and higher density compared with the **Paraffin** compounds. The average crude oil contains about 50 per cent by weight of **Naphthenes**. In general, heavy oils have higher proportions of **Naphthenic** compounds.

The Cycloalkenes Series

Like the **Cycloalkane**, there exist the **Cycloalkene** series which are unsaturated compounds of molecular structural form that contain one double carbon-to-carbon bonds. The **Cyclobutene** is an example for this type of series, shown as follows:

Cyclobutane (C_4H_8)

```
    H   H
    |   |
H - C - C - H
    |   |
H - C = C - H
```

Example of the unsaturated ring-form hydrocarbon Cycloalkene, Cyclobutene (C_4H_6).

The Aromatic Hydrocarbons

These hydrocarbons (also called Arenes) are named as Aromatic because many of these compounds have sweet and pleasant odor. The simplest form in this type of hydrocarbons is Benzene, (C_6H_6) which is an unsaturated single ring compound made up of six carbon atoms, with alternating double and single bonds existing between carbon atoms.

```
        H
        |
        C
      //  \
H -- C      C -- H
     |      ||
H -- C      C -- H
      \\  /
        C
        |
        H
```

Single ring aromatic hydrocarbon, Benzene (C_6H_6).

Benzene is considered to be an important industrial substance.

1.3 Hydrocarbon Deposits in Nature

In nature hydrocarbon deposits exist in various physical states ranging from the solid primitive immature petroleum matter (Kerogen) to the liquid and gaseous state petroleum deposits. An explanatory note is given here-below on each of these matters.

1.3.1 Kerogen

Kerogen is the term used for the primitive hydrocarbon matter from which petroleum and hydrocarbon gases are originated. It consists of fine-grained, amorphous matter which is insoluble in the usual oil-solvents like carbon disulphide, for example. On average, Kerogen consists of carbon (75%), hydrogen (10%), and the rest (15%) is of other chemical elements like sulfur, oxygen, and nitrogen (Selley, 1983, P. 1).

It is to be noted that, when the ratio of Kerogen exceeds 10% of rocks (in weight) it becomes of commercial significance. In this case, crude oil can be recovered from the Kerogen-bearing rocks by heating process. Produced oil (from oil-shale and from oil-sand) can be recovered in this way from certain shale- or sandstone-rocks.

1.3.2 Asphalt

Asphalt is hydrocarbon in its solid state at normal pressure and temperature. Unlike kerogen, Asphalt is solvable in the usual oil solvents. The Asphalt matter deposits are generated from incomplete thermal maturation process of Kerogen, or from crude oils from which gases have been dissipated as a result of oil accumulation that occurred in improperly closed oil-traps.

1.3.3 Crude Oil

Crude oil is the hydrocarbon deposits which are of liquid state at normal atmospheric conditions, and it is soluble in oil-solvents.

1.3.4 Natural Gas

Natural gas is the hydrocarbon gas associated with the petroleum accumulations. It exists either as gas dissolved in the oil body or as a separate body of free-gas accumulation. Natural gas consists mainly of methane, ethane, propane, and butane. In addition to these hydrocarbon gases, natural gas contains small proportions of other gases of non-hydrocarbon types. In particular, it contains hydrogen sulphide, (H_2S) in which case the gas is described as "sour gas" to discriminate it from the gas that is void of the hydrogen sulphide which is called "sweet gas".

1.3.5 Natural-Gas Condensates

Natural-gas condensate is a low-density (API degree >50) mixture of hydrocarbon liquids which usually occurs in association with natural gas. The produced hydrocarbon vapor gets condensed, that is transformed into liquid state, when temperature is reduced to a level below the dew-point temperature of that gaseous hydrocarbon. The condensate hydrocarbon is considered as light crude oil of S.G.(0.5 to 0.8) under normal atmospheric pressure when brought up to the earth surface.

In general, condensates are composed of paraffin compounds like propane, butane, pentane, hexane, etc. In addition to that, condensates contain impurities such as hydrogen sulphide and carbon dioxide.

1.4 The Crude Oil

The crude oil is that hydrocarbon substance, which is naturally occurring in the subsurface rock media. It is liquid at normal pressure and temperature conditions. In other words, it is the oil that flows out from an oil-well. In general, crude-oil density fall roughly in the range of (0.8 – 0.9) gm/cc. Both of the density and viscosity are normally less when the crude oil is under reservoir conditions and that is due to the relatively high temperature and due to the increase in the gas dissolved in the trapped oil body.

1.4.1 Crude Oil Composition

Crude oil is mixture of hydrocarbon compounds covering the whole range of hydrocarbon series (simple straight, branched, and cyclic chains). These series are mostly of saturated types. Unsaturated hydrocarbons, such as alkenes, cycloalkenes, and aromatics are present in very small quantities. The crude oil contains other inorganic compounds that include the main chemical elements (Nitrogen, Oxygen, and Sulfur, NOS) in addition to traces of metals, or metal impurities. The most abundant trace-metals are Vanadium and Nickel. It is common that crude oil contains the inorganic compound hydrogen sulfide, which exists either as free or as dissolved gas. Crude oil differs in composition from reservoir to reservoir in the same oil field and from oil field to oil field. However, it is found that, on average, it is composed of the carbon element by a ratio in the range (82% - 87%) and hydrogen (12% - 15%) plus the three elements (NOS) in a proportion of less than (5%), (Chapman, 1976, p. 28).

Crude oil containing more than 6% of sulfur is called "sour crude oil". Presence of high proportions of sulfur causes degradation to the oil quality, in which case extra refining processes have to be carried out in order to remove the sulfur impurity. Oils having lower sulphur content are called "sweet crude oil".

1.4.2 Types of Crude Oil

Depending on the proportions of the different hydrocarbon constituents, crude oils are classified into different types. For example, the oil that contains high proportion of paraffinic hydrocarbons is called "paraffin-base oil". It is, however, called "mixed-base oil" in case it is made up of nearly-equal proportions of the different hydrocarbons. The naphthene-base oils are often made up of mixture of large-molecule hydrocarbons, in addition to other compounds that have the NOS elements with little amounts of trace metals as Nickel and Vanadium. Crude oils consisting of large-molecule hydrocarbons are called asphalt-base oils.

Based on these basis, crude oils are divided into three types. These are:

(i) Paraffin-base oil, most of which is consisting of paraffin-series.
(ii) Asphalt-base oil, most of which is consisting of naphthene-series.
(iii) Mixed-base oil, most of which consist of nearly equal proportions of paraffin and asphalt components.

The principal difference between paraffinic and asphaltic oils is that the paraffinic oils are generally lower density and lighter in colour compared with the asphaltic oils. In all of these types of crude oils non-hydrocarbon compounds are found. It is noted that the distillation fraction compounds contain greater proportions of these non-hydrocarbons, high molecular weight NOS-compounds. These are commonly found in the asphaltenes or resins material produced by oil-refineries.

1.4.3 Effects of Geology on Crude Oil

Since oil matter is created and deposited within the pores of subsurface rocks it is expected that the geological nature of host medium has an effect on the type and composition of the oil accumulations. In practice, it is found that rocks which contain sulfur in their composition (gypsum, anhydrite, and calcareous rocks) produce sulfur-bearing oils, whereas a rock medium made up of sand-shale sequences would have paraffin-base oils which are usually sulfur-free or have very small amount of sulfur. The oil provinces, in the Northern parts of Iraq for example, where limestone and anhydrite are dominant, oil deposits are of asphaltic type with appreciable sulfur ratio. The oil in Sudan where the reservoir rocks are mainly of sand-shale sequences, are typical example of the sulfur-free waxy or paraffinic type.

It has been reported that there are hydrocarbon gases which are rich in non-hydrocarbon compounds. Some of these hydrocarbons are found to be rich in compounds like carbon dioxide (CO_2) and Nitrogen (N_2) gases. An example of such cases is the gas produced from some of German off-shore sources. These hydrocarbons are believed to be associated with deep-seated igneous intrusions or with other types of deep-rooted geological structures (Hobson, 1984, p 39).

1.5 The Natural Gas

Natural gas is that part of hydrocarbon matter which is found in gaseous state. It exists either as gas dissolved in the liquid or as separate free-gas collected over the oil accumulation in a form of a gas cap. By application of pressure, natural gas can be liquefied as it is done when processed and filled in cylinders made available for domestic uses. Natural gas is more often found in subsurface oil traps, associated with crude oil in the form of gas-caps.

The associated gas which gets separated from the crude oil, as part of the production process, can be processed to be used as energy source, or re-injected in the reservoir to enhance oil recovery. The gas can also be used as hydrocarbon source for producing petrochemicals. Gas flaring is a common practice followed by oil producers to get rid of the unused quantities in order to avoid its pollutant effects on environments.

1.5.1 Natural Gas Composition

The associated gas consists mainly of methane gas (CH_4) with small percentages of the other paraffinic gases like ethane (C_2H_6), propane (C_3H_8), and butane (C_4H_{10}). These four hydrocarbons exist in gaseous state at surface normal pressure and temperature (Stoneley, 1995, p 29). Associated gas contains also vapors of other hydrocarbons, of quantities depending on the reservoir conditions. Normally, natural gas also contains small percentages of non-hydrocarbon compounds, most of which are nitrogen (N_2), carbon monoxide (CO), carbon dioxide (C_2O), and hydrogen sulfide (H_2S).

1.5.2 Types of Natural Gas

Natural gas can exist as dissolved in the crude oil or as separate in the form of a gas-cap formed over the liquid crude oil present in subsurface oil-traps. The raw natural gas that is produced from a crude-oil producing well, is usually referred to as the (associated gas). The non-associated gas is produced from gas reservoirs which contain only free gas with no any hydrocarbon liquids present with it.

Associated Natural Gas

The associated natural gas is originally formed as hydrocarbon gases dissolved in the crude oil by quantities governed by the reservoir pressure and temperature conditions. When the pressure inside the oil reservoir (called reservoir saturation pressure) is at a level at which dissolved gas starts to liberate from the gas-saturated oil, a gas-cap is created. In this case the oil reservoir is called (saturated reservoir). If, however, all gas contents are dissolved in the liquid oil, no gas-cap is formed, and, in this case the reservoir is referred to as an (under-saturated reservoir).

Sweet and Sour Natural Gas

Natural Gas is described as "sweet" or "sour" according to the amount of hydrogen sulfide (H_2S) present in the gas body. It is called "sweet gas" in case of low-level H_2S percentage, and "sour gas" when the percentage is high. The proportion of the hydrogen sulfide, which defines these two types of gases, is not fixed. Different companies adopt different values of percentages. However, the hydrogen sulfide present in the natural gas is used as the yardstick for differentiation between "sweet" and "sour" gas.

Dry and Wet Natural Gas

The gas is said to be "dry gas" when it is almost completely made up of methane. Normally, dry gas occurs in gas-reservoirs, where the whole hydrocarbon reservoir is free gas.

When the gas is containing appreciable proportions of hydrocarbon matter of molecular structures having number of carbon atoms greater than one (i.e. larger molecules than that of methane), the gas is called "wet gas". Typical paraffin series composing the wet gas are ethane, propane, and butane. The heavier hydrocarbon-gases get condensed when brought to the surface in which case it can be separated as light hydrocarbon liquids normally called (natural gas liquids, NGL). The lighter gases (propane and lighter series), on the other hand, can be industrially liquefied to form what is commonly called (liquefied petroleum gas, LPG).

1.6 The Formation Water

A typical oil reservoir is made up of three distinct layers of liquids. Oil layer in the middle with gas cap above it and water body at its base. The water naturally existing in the pores of rock formations, just below the oil accumulation is called (formation water) or the (interstitial water). This is to be differentiated from the fossil water (connate water) which is the water trapped in the rock pores that took place during its formation. In an oil reservoir the formation water exists as a separate free-water layer existing underneath the oil zone. Noteworthy, is that, minute quantities of the formation water are trapped in the pores that contain oil and gas, that is in the oil and gas zones.

1.6.1 Types of Salt Contents

In general, formation water is highly saline due to its salt contents of various types. Typically, salt contents comprise of sodium chloride (NaCl) with less proportions of sodium sulphate (Na_2SO_4), potassium sulfate (K_2SO_4), calcium chloride ($CaCl_2$), calcium sulfate ($CaSO_4$), and magnesium sulfate ($MgSO_4$). Water salinity is due mainly to soluble salts, as the chloride salts, that are dissolved in the moving water during fluids migration from the source rocks to the reservoir rocks where accumulation takes place.

1.6.2 Role of Geological Conditions

The water salinity varies from reservoir to the other depending on the migration and accumulation conditions. The governing conditions are the geological nature of rocks and pressure and temperature conditions of the containing reservoir. In practice, it is found that a direct relationship exists between type of crude oil and the salinity of the associated formation water. An inverse relationship is reported to occur between oil density and water salinity. Also it has been observed that there is a direct relationship between the percentage of the sulfate salts dissolved in the water and the crude-oil sulfur content. This is supported by the general finding of the high sulfur content in the oil produced from carbonatic reservoirs compared with oils from sandstone reservoirs.

Studies of variations of the salinity degree have important reservoir-engineering applications. Such studies can lead to useful information as regards oil-migration directions, and hence, to location of the source rocks. Another important application of the formation-water behavior is study of the water-pressure variations in different parts of a reservoir. These variations would give indications on the rocks permeability barriers and faulting patterns. Thus, the case where the pressure was found to be different from the normal hydrostatic pressure can indicate changes in geological structure of the reservoir.

1.7 Oil Refining and the Refinery Outputs

The crude oil that comes out of the oil well is mixture of different hydrocarbon compounds including dissolved organic and non-organic gases, in addition to water and rock minute particles. In normal oil production, crude oil is moved out of the oil well, heading for the export terminal or to an oil refining plant. At a point between the well and the refinery the oil is passed through a special oil-production installation system, called a (separator) or (degassing station as it is sometimes called). This installation is designed to separate the main three components, namely the oil, gas, and water. This, partially purified crude oil, represents the basic input to the oil-refinery complex in which the refining processes are executed. The oil refining process is considered as the main phase of the "downstream" activities of the petroleum industry.

1.7.1 The Oil Refinery

An oil refinery is an oil-processing plant which outputs separate refined hydrocarbon products from the input crude oil. The separation process is accomplished through a giant cylindrical tower called (oil-fractionating tower). In this tower, the hydrocarbon components are separated into the "fractions" which are making up the input crude oil. The fractionating towers form the prominent features of all oil-refinery installations.

The fractionating tower is a giant steel cylinder of about (12 to 24) feet diameter with height reaching (100 to 150) feet (Wymer, 1964, p. 65). Inside the tower cylinder there are many horizontal layers of trays with holes in them. When at work, the tray-layers are so designed that each layer would be at a certain temperature different from the adjacent tray-layers. The temperature is controlled in such a way as to have the highest temperature is at the lowest level in the tower, whereas the lowest temperature is at its top. The holes in the trays are made so that the heated hydrocarbon vapors would pass through them. The tray-holes are provided with basin-shaped covers for the purpose of forcing the vapors rising through the holes to 'bubble out" through the liquid collected on the tray.

The crude-oil components that have highest boiling points get condensed on the lower trays and those with the lowest boiling points continue moving upwards towards the top of the tower where they get condensed on the relatively colder trays. Outlets from these trays allow the tray-content to be taken away for storage. It should be noted here that in the one refinery there may be several fractionating towers in operation at the same time.

A schematic representation of the fractionating tower is shown in (Fig. 1.4).

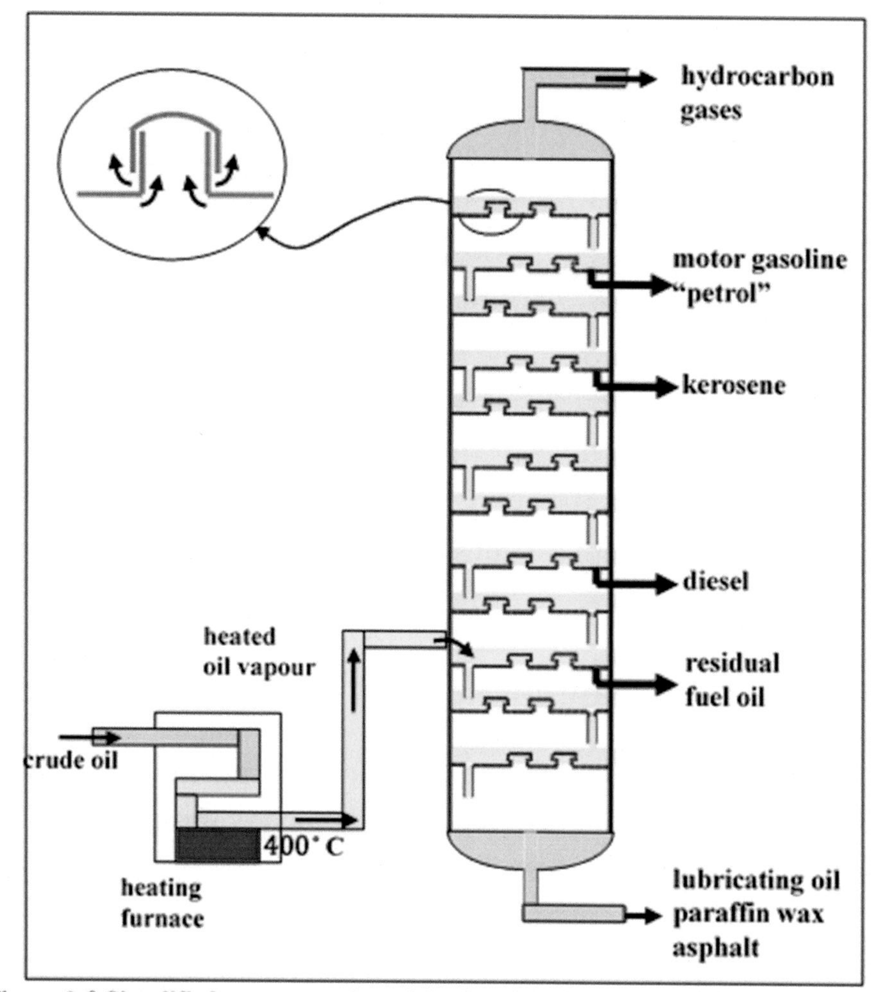

Figure 1-3 Simplified representation of the fractionating tower and its tray-layers which are designed to collect the separated refined products.

In addition to the fractionating tower, the refinery may be equipped with other specialized units designed to perform some product-upgrading processes by chemical techniques. Oil refineries are usually equipped with a tank farm for receiving and storing the incoming crude oil which is ready to be fed to the fractionating tower for the distillation process. Some of the tanks in the tank farm are used to store the outputted refined hydrocarbon products.

Usually oil refineries are built by water areas such as sea coasts, rivers, or water canals. This is because of the great amounts of water needed in machinery operation. Also, because of the convenient transport-means, water-ways provide for both crude oil supplying and refined-products marketing.

1.7.2 The Refining Processes

In general the input to the refinery complex is the crude oil which itself an output from a previous preliminary purification system; the degassing station. In addition to the crude oil, other certain types of oils, such as partly-refined hydrocarbons, may be inputted to undergo further refining processes.

The basic principle, upon which the refining processes depend, is the (fractional distillation). According to this principle, the components of the hydrocarbon mixture of the input crude oil, are sorted out and separated according to the boiling point of each fraction. The main products of the refining process therefore include, but not limited to, gasoline, kerosene, and liquefied petroleum gas, diesel fuels lubricating oils, asphalt, and other heavy hydrocarbon residues.

The distillation process starts by heating the incoming crude oil to a high degree (about 400 degree Celsius) then it is pumped through the fractionating tower, in which the hydrocarbon components are separated into the "fractions" which are making up the input crude oil. The separation process is achieved according to the boiling point of each of the hydrocarbon components. At the base of the tower, the heavy hydrocarbons (as tar and other heavy residues) are collected, whereas the gas and other light products are formed in the upper part of the tower.

1.7.3 The Refinery Principal Outputs

As it is mentioned above, refined hydrocarbon fractions outputted from the fractionating tower are separated in accordance with their individual boiling point. Starting with the lightest hydrocarbons produced from the top of the tower, the principal outputs are the following:

- Hydrocarbon gases:

At the top of the tower, hydrocarbon gases, consisting mainly of Butane and propane, are produced. These can be passed into a condensing system to produce the liquefied petroleum gas (LPG). Special gas cylinders can be filled in with the produced LPG to be used for domestic applications.

- Motor gasoline (Petrol):

Motor gasoline, commonly known by the name (car petrol, or just petrol), is considered to be the most important refinery product which is further treated to be of quality and properties suited for motor engine best-performance. For this purpose, antiknock feature (rated by a measure called the octane number) and protection against metal rusting, are secured by chemical treatments and specially-made chemical additives.

- Kerosene

Kerosene and diesel fuel are middle-distillate products which are important in their wide-scope uses. Kerosene is used as a jet fuel, domestic heating, and other less important uses as in cooking, lighting, and as a solvent.

- Diesel

. Diesel is the well-known motor fuel is extensively used in transport machineries as in heavy trucks, buses, and trains.

- Heavy fuel oil

Heavy fuel oils (also called residual fuels) are the remaining hydrocarbons after the distilled fuel oils and the other lighter hydrocarbon products are distilled away. These are used as energy source for driving power plants, industrial machines, and ships.

- Lubricating oil, Paraffin waxes, - Asphalt

Lubricating oils, paraffin waxes, and asphalt are the hydrocarbon products formed at the bottom of the fractionating tower. These materials are characterized by their high viscosity indexes and have valuable industrial applications.

1.8 The Petrochemicals

It can be said that there are two approaches followed in deriving useful products from crude oil. These are based on physical and chemical methods of approach:

The physical method is represented by the fractional distillation method (described above) which is based on heating the crude oil and then separating the various components according to their individual boiling points. This method leads primarily to production of mainly oil fuels of various types.

1.8.1 Definition of Petrochemistry

The chemical methods, on the other hand, are based on chemical reactions that lead to production of new chemically-derived materials which are collectively called (Petrochemicals). These materials are named petrochemicals, indicating that they are petroleum-products which are derived by chemical means A branch of chemistry that is concerned with deriving new substances from hydrocarbon compounds by chemical reactions is called (Petrochemistry).

Since the early years of the twentieth century, petrochemical industry has flourished, and petrochemical products have played a great role in domestic and industrial activities. To give an idea about the present size of petrochemical industry, it is estimated that the annual world-production is about (170) million ton of ethylene, (125) million ton of methanol and (120) million ton of propylene, (Eramo, 2015).

The three hydrocarbons (ethylene, propylene, and aromatics) are considered to be the basic raw materials from which a great number of other petrochemical products are chemically manufactured.

1.8.2 The Chemical Means of Approach

Two main types of approach are generally followed in manufacturing refined hydrocarbon products by chemical reactions. These two approaches are chemical composing and decomposing processes of hydrocarbon molecules. The raw material for the process is either light hydrocarbons (as natural gas) or heavy hydrocarbon matters as those outputted from the lower part of the fractionation tower of a refinery.

The common procedure in molecule-composing, is a chemical process called (polymerization), which is a process by which many hydrocarbon molecules (i.e. many monomers) are bonded together to form a new poly-molecule compound (polymer) with its own physical and chemical properties. The other type of approach (chemical molecule-decomposing) is breaking up (cracking) of the large hydrocarbon-molecule into smaller molecules by use of a special steel container (cracker) in which a hydrocarbon-product (such as gasoline) is circulated at high temperature with the presence of a catalyst. With this process (catalytic cracking) a heavy refined gasoline (petrol) can be converted into high-grade hydrocarbon product.

1.8.3 The Raw Materials for Petrochemical Products

The most common hydrocarbon compounds which are used as raw materials for making petrochemicals are the unsaturated paraffin series, the olefins which is an alkenes series. The principal hydrocarbons used, are olefins (including mainly ethylene and propylene) and the aromatic series (including benzene and toluene). Both of the olefins and aromatics are produced by oil refineries through catalytic cracking of oil fractions. At a later stage, these materials (olefins and aromatics) are converted into materials of more direct use by consumers. For example, the ethylene is converted into the plastic product, which is commonly known as (polyethylene). Also, benzene is converted into the material (nylon) which is used in textile and clothing industry. In fact, both olefins and aromatics serve as the raw materials for a wide range of petrochemical products, such as plastics, solvents, detergents, artificial rubber and many more.

1.8.4 Role of Polymers in the Petrochemical Industry

The polymer (ethylene) has an exceptionally active role in the petrochemical industry. It can be obtained from a paraffin matter by a catalytic cracking process. Thus, for example, when propane (C_3H_8) is heated to a high temperature with presence of catalyst, it yields ethylene (C_2H_4) plus methane (CH_4), that is:

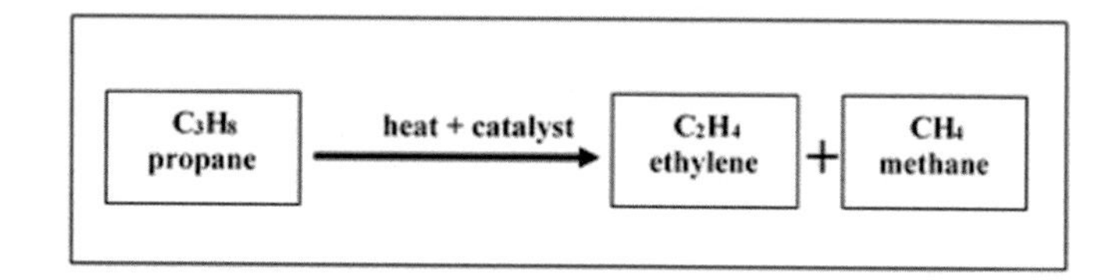

With the polymerization process, ethylene can be converted into the petrochemical plastic substance, the (polyethylene).

It should be noted here that the most widely produced polymers are polyethylene, polypropylene, and polyvinyl chloride (PVC). The polymer PVC is largely used in making pipes, doors, and windows. The total global production of PVC reached the level of 24.9 million ton in the year 1998 (Stringer, 2002, p. 79).

1.8.5 Examples of Petrochemical Products

Petrochemical materials are now covering a very wide range of applications in both domestic and large-scale industrial fields. Some of these petrochemicals are soft (as Cellophane sheets) and others are hard (as Perspex).

Examples of petrochemical products are: synthetic fibers, synthetic rubber, electrical insulators, data-storage media (plastic magnetic tapes and compact discs), dyes, fertilizers, insecticides, soaps, detergents, and many more.

1.9 Physical Properties of Crude Oil

Crude oil differs in physical properties depending on the geological environment within which the oil has been generated, migrated, and finally accumulated. Oil may be considered (light oil) or (heavy oil) depending on its density. Further, it may be called sweet or sour depending on the sulfur content. A number of physical properties are normally employed in describing a crude oil. Most important properties are the density, viscosity, surface tension, flash point, refraction index, caloric value, and electrical conduction. Definition notes on these properties are here-below presented.

1.9.1 Density

Crude oils are generally classified into light (low-density oil), medium, and heavy (high-density oils), according to density. This property reflects the nature of hydrocarbon components of the crude oil. In consequence, oil density is used as the principal rating factor, upon which oil pricing is determined.

Density rating may be evaluated by an absolute value by the well-known units (gm/cc) or by use of the relative measurement-system using the unit-less (specific gravity, SG) method. Since, volume is temperature-dependant; density must always be measured at a fixed temperature vale. The specific gravity of any substance is defined as the ratio of the mass of a substance to the mass of the same volume of water measured at a defined temperature. However, density of crude oils is usually measured by another especially defined density-scale. This is the API-scale which is devised by the American Petroleum Institute. The API-degree which is calculated by an SG-dependant formula of the following form:

API degree = 141.5/SG (at 60°F) – 131.5,

and,

SG (at 60°F) = 141.5/(API degree + 131.5)

These formulae show that, the API degree (or API gravity, as it is often called) is inversely proportional to SG. Also, these show that the API of water (SG=1) is 10 and oils which are lighter than water will have API degrees greater than 10. In this way API becomes a measure of heaviness and lightness of oil compared with water density. Thus, oil is considered to be light if its API degree is greater than 10, and heavy if its API is smaller than 10.

The mathematical relationship between API degree and SG is shown in the following chart (Fig 1-4).

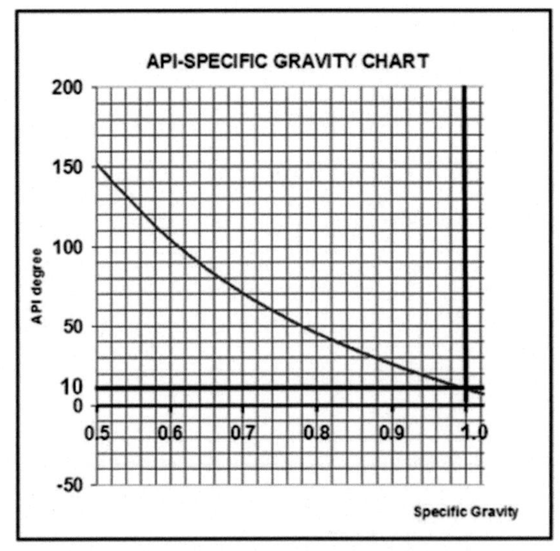

Figure 1-4 Chart of the API-SG relationship.
The point (1.0, 10) represents the water point on the chart.

This chart can be used in conversion of SG values into API degrees and vice versa.

Based on the API degree, oils of API values within the range (10 - 20) are considered to be heavy, whereas those oils having API greater than 30 are considered to be light. Typically, light crude oils are of API degrees which fall in the range (30 - 50). Medium oil has API degree within the range (20 - 30), while oils of API below 10 are considered to be extra-heavy oil.

This simplified measurement-scheme is used in classification of crude-oils as shown in (Fig 1-5).

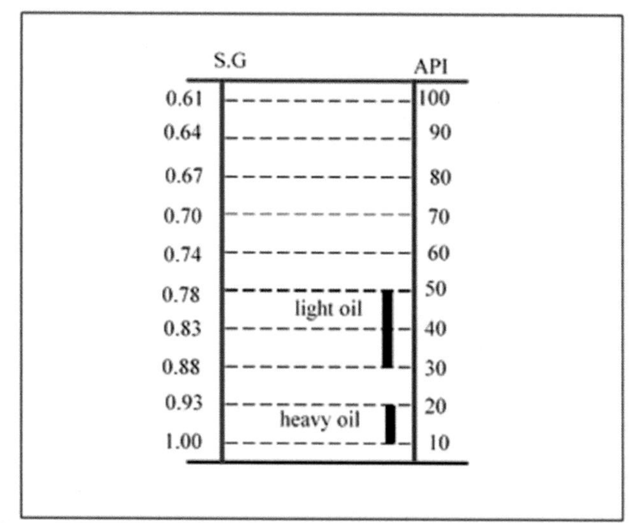

Figure 1-5 Classification of crude oils on the basis of API degree

It is useful to note that crude-oil quantities are generally measured in barrels, where one barrel weighs roughly about 300 pounds. The exact weight of a barrel depends on the API degree of the particular oil. Another common measure is the metric ton which is equivalent to about 7 barrels.

1.9.2 Boiling-, Freezing-, and Pour-Points

When a liquid is heated, a temperature is reached when the vapor pressure reaches a level equal to that of the surrounding pressure. At this point, bubbles are being formed inside the liquid body which is described as boiling liquid, and the temperature at which a liquid boils is called the (boiling point). Again, any liquid has a characteristic temperature at which it would turn into solid state. The temperature at which a liquid is converted into solid is called (freezing point). There is another characteristic temperature associated with liquids and that is the (pour point) which is defined to be the temperature below which the oil becomes semi-solid (plastic nature), at which the liquid cannot flow freely.

Because crude oil is made up of a mixture of many types of hydrocarbon compounds, it is not possible to assign one boiling point (or a freezing point) to crude oil. In fact each hydrocarbon-component has its own boiling (or freezing) point. As regards the pour-point parameter, a crude oil can have a pour-point which varies as its type and its component percentages. Thus, for example, an oil of high paraffinic components would have a high pour-point.

1.9.3 Viscosity

Viscosity of a fluid is an expression for its resistance to flow. Difficulty of liquid flowing is due to the internal friction existing within the liquid body. It is measured by a unit called (Poise), where (1 Poise = 1 dyne.sec.cm^{-2}).

In general, viscosity of a crude oil varies directly with temperature, and inversely with the dissolved gases within the oil liquid. It is especially important in giving useful information as to the components of a particular crude oil. Thus, from the viscosity-density combination, valuable information can be inferred as regards composition of the crude oil. For example, lower viscosity-density values, is indicative of increase of paraffinic fractions.

1.9.4 Surface Tension

It is a common observation that the free surface of any liquid behaves as if it is under tension. This phenomenon (termed surface tension) is due to inter-molecular attractive forces (cohesion forces) binding the molecules forming the free surface of the liquid body. Since there are no molecules above the surface, a net inward force is created at that surface causing it to contract, resisting any stretching caused by other sources. The net result is that the surface will behave as if it is covered by stretched elastic membrane or "skin". Surface tension is measured in (dyne/cm) or in (erg/cm^2).

Surface tension varies with the type of liquid and with temperature for a given liquid. As temperature of a liquid increases, its surface tension decreases, and for the same temperature, surface tension of a hydrocarbon is greater the higher the molecular weight. Together with viscosity, surface tension has an important role in the oil-recovery process.

1.9.5 Flash and Burning Points

The (flash point) is the lowest temperature at which vapor of a hydrocarbon fluid, in air, will get ignited by an ignition source, without continuing to burn. At the flash point burning stops once the ignition source is removed. With increase of temperature beyond the flash point, a temperature is reached at which ignition will start continuous burning. This temperature is called the (burning point) or (fire point).

The flash and burning points expresses the easiness of a certain substance (such as lubricant) to get ignited and burned. Thus, materials having low flash points are more flammable and hence of higher fire-risks. For this reason they become useful parameters which are taken in consideration in designing transportation and storage facilities.

It should be noted that the auto-ignition temperature is a temperature at which the hydrocarbon substance will ignite due to heating (temperature increasing) only without presence of an ignition source such as a spark or a flame.

1.9.6 Refractive Index

A ray of light which is obliquely incident on an interface separating two media, will bend (get refracted) on passing from one medium into the other. The refracted ray will be bent either towards, or away from, the normal line. For any two adjacent media, the angle of refraction is function of the angle of incidence as well as of the velocities of the moving wave in the two media. The mathematical relationship (called Snell's Law) governing the incidence and refraction angles (θ_1, θ_2) with the velocities ($\mathbf{v}_1$, $\mathbf{v}_2$) of the two concerned media is given by ($\sin \boldsymbol{\theta}_1 / \sin \boldsymbol{\theta}_2 = \mathbf{v}_1 / \mathbf{v}_2$). In the physics of optics, the refractive index, also called (index of refraction), is defined as the ratio ($\mathbf{v}_1/\mathbf{v}_2$), where ($\mathbf{v}_1$ and $\mathbf{v}_2$) represent the speeds of a monochromatic light (light of a given wavelength) in vacuum and in the substance respectively. In other words, refractive index of a substance gives a measure of the speed of light, of a given wavelength, in that substance. Since light travels in vacuum with the highest speed than in any material, refractive index of any substance is greater than one. For example refractive index of water is (1.33).

The refractive index can be used in characterization of crude oils. As with density, low values of the refractive index are associated with paraffins, and higher values with aromatics (Banks and King,1986, p. 322).

1.9.7 Specific Heat

Specific heat of a substance is defined to be the number of calories required to raise the temperature of one gram of that substance by one degree centigrade. Specific heat varies directly with temperature and inversely with the density.

1.9.8 Electrical Conduction

All hydrocarbons (crude oils and their fractions) are characterized in being of poor electrical conductivity. For this reason, petroleum-derived materials are used for electric insulation.

Chapter-2

2. GEOLOGICAL BACKGROUND

World-wide oil-exploration efforts are being extensively spent to find hydrocarbon accumulations, which are normally trapped at depths of few kilometers below surface. Naturally, generation of the hydrocarbon matter, its migration from the source rocks, and entrapment within reservoir rocks are all governed by the prevailing geological conditions. In view of the strong ties connecting petroleum deposits with the overall geological environments, giving an overview of some geological concepts would be beneficial in understanding the subsurface oil behavior. Accordingly, a brief review on the geological aspects of petroleum deposits will be presented in this chapter.

2.1 The Earth Science

The science of geology, which deals with rocks of the Earth crust, is part of an ensemble of interrelated geological sciences which are collectively referred to as the Earth Science (also called Geoscience). Earth Science includes Geology and its sub-geological divisions as, Geophysics, Geochemistry, and several other related sub-divisions.

The Earth and natural activities occurring in, and over its surface, may be represented by a simplified model that consists of three parts; matter-element, energy-element, and biological-element (animals and plants). This model allows dividing the science involving these activities into the three basic sciences which are Chemistry (for matter composition), Physics (for types of energy), and Biology (for animals and plants). The Earth Science (commonly referred to as Geology) can be seen to be occupying a central position within these three basic sciences, giving rise to three principal sub-branches of science; Geochemistry, Geophysics, and "Geobiology" that includes Biostratigraphy and Palaeontology (Fig. 2.1).

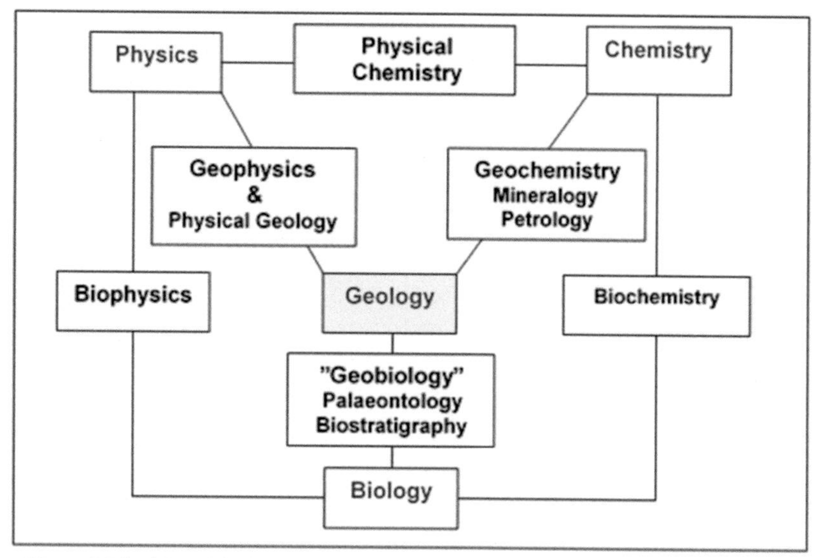

Figure 2.1 Position of Geology within the three basic natural sciences, Chemistry, Physics, and Biology.

2.2 Geology Main Subdivisions

In the literature of history of sciences, it is stated that modern geology began in the medieval Islamic world, by Abu Al-Rayhan Al-Biruni (973-1048 CE). Another Islamic scholar is Ibn Sina, who, at that time, proposed detailed explanations for the formation of mountains, and other topics which are considered as the foundations for the later developments of Geology. The more recent advances were the outcome of international joint effort from scholars from all parts of the world. The two British scholars, William Smith (1769-1839) and James Hutton (1726-1797) are now considered to be pioneers of modern Geology.

Geology is considered to be one of the main pure sciences that witnessed extensive developments in its theoretical concepts and in its scientific scope of coverage. Its principal concern is the study of the crustal rocks of the planet Earth. Geological research activities were further extended to cover structural features of other planets and natural satellites as the geology of Mars and the Moon.

The main subdivisions of geology are the following:

(i) Physical geology – Geological processes as rock deposition, erosion, and weathering
(ii) Historical geology (Stratigraphy) – Historical developments of rock strata
(iii) Petrology and Mineralogy – Rock types, compositions and alterations
(iv) Geophysics – Physics of the Earth potential and other natural fields
(v) Geochemistry – Chemical properties and changes occurring in the crustal rocks
(vi) Structural geology – Geometrical shapes of strata including folding and faulting.
(vii) Palaeontology – Biological remains (fossils) preserved in rocks.

(viii) Mineralogy and Crystallography - Mineral genesis, types, and crystal forms

Basically, Geology is a pure science, but geological principles have been put to applications by devising appropriate technologies and engineering methods. Examples of applied-geology fields are the following:

(i) Petroleum geology – Evaluation of geological environments of petroleum deposits
(ii) Economic geology – Feasibility assessments of oil and mineral resources
(iii) Mining geology – Study of the techniques of extracting raw mineral-deposits
(iv) Engineering geology– Application of geologic principles for engineering purposes
(v) Hydrology – Evaluation of geological environments of water resources

2.3 Petroleum Geology

Petroleum Geology is concerned with the overall geological environments of petroleum including its behavior throughout its phases of development; generation, migration, and accumulation (entrapment). One of the principal goals of petroleum geological studies is determination of the geological model of the oil field under study. The model is continuously updated when additional geological, geochemical, and geophysical data are accumulated. In practice, investigations are directed towards prediction of the characteristics of the oil-bearing rock formations and their potential hydrocarbon contents.

Accuracy of the determined geological model of the studied reservoir, and its close resemblance to the actual geological structure is very important issue, since all the following development-activities are based on the predicted model. In particular, drill-hole locations, drilling parameters, production and field-development activities, which are costly processes, must be optimum in design and parameters.

Petroleum geological studies lead to determination of the appropriate engineering tools applied to get optimum outcome of an oil field.

2.4 The Planet Earth

The Earth is not a perfect spherical body, but it is of an ellipsoidal shape, which has its polar radius less than the equatorial radius by 21 km. This shape is believed to be resulting from the centrifugal force created as result of the Earth rotation about its polar axis. The main statistics of the planet Earth is quoted in (Fig. 2.2).

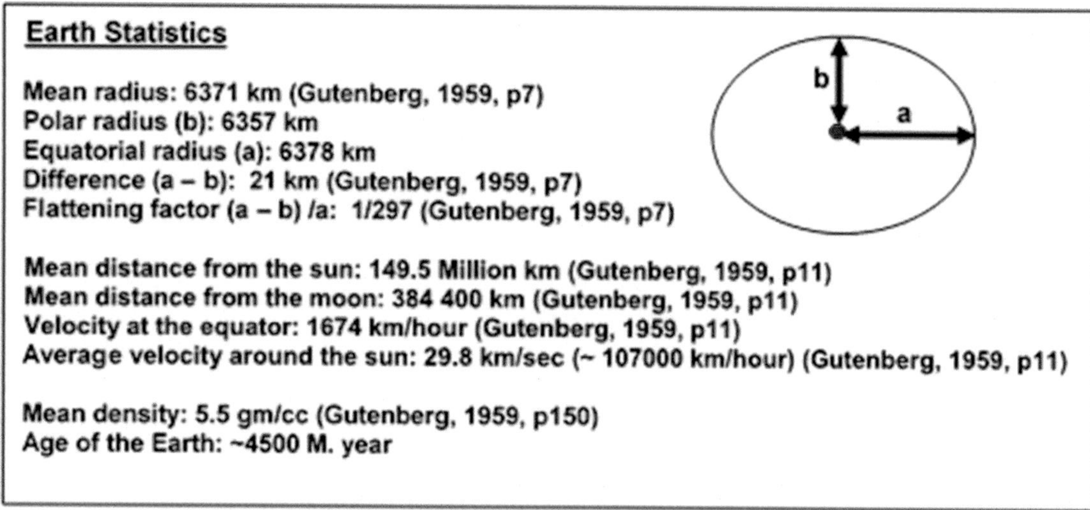

Figure 2.2 Main statistics of the planet Earth (Gutenberg, 1959).

2.4.1 Structure of the Earth

The planet Earth consists of three main zones; the Crust, the Mantle, and the Core. The core is further subdivided into the liquid Upper Core and solid Inner Core. The discontinuities separating these zones are called, the (Mohorovicic or Moho discontinuity) for the Crust/Mantle discontinuity, the (Gutenberg discontinuity) for the Mantle/Outer Core discontinuity, and the (Lehman discontinuity) for the Outer Core/Inner Core discontinuity. Each of these discontinuities represents a sudden change in the value of the P-wave velocity, indicating change in the physical properties between the zones of the Earth interior (Fig. 2.3).

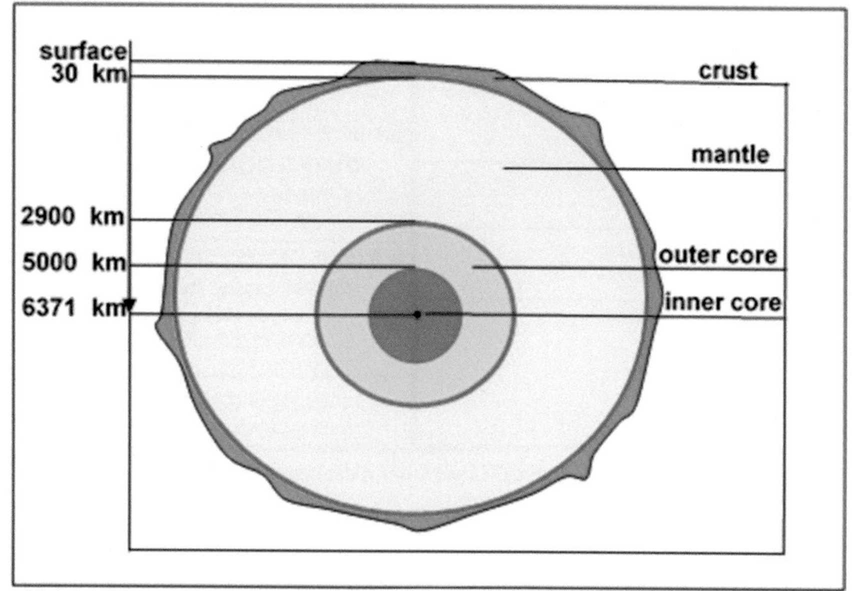

Figure 2.3 Structure of the Earth interior

Determination of the Earth internal divisions were based on the studies of the seismic waves which are generated by earthquakes. The Core was deduced from existence of the shadow zone which was interpreted as being formed by an existing spherical body (the Earth Core) which is acting as a spherical lens that prevents incoming waves from reaching the Earth surface opposite to the earthquake focus (Fig. 2.4).

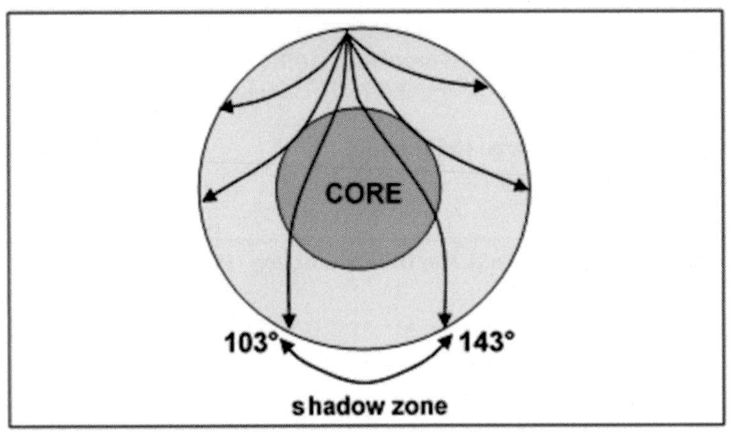

Figure 2.4 Observation of a shadow-zone of earthquake wave arrivals led to the discovery of the Earth Core

The Earth zones and discontinuities separating them are summarized in the following figure (Fig. 2.5).

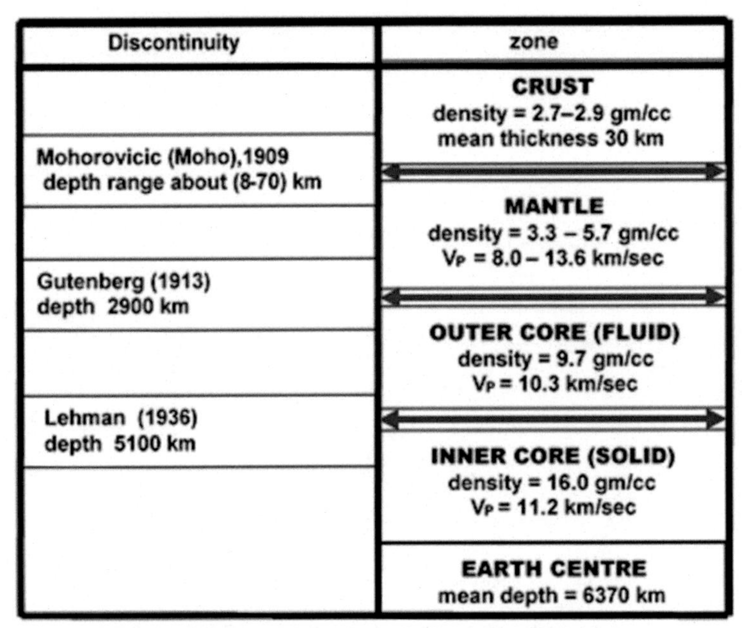

Discontinuity	zone
	CRUST density = 2.7–2.9 gm/cc mean thickness 30 km
Mohorovicic (Moho),1909 depth range about (8-70) km	
	MANTLE density = 3.3 – 5.7 gm/cc V_P = 8.0 – 13.6 km/sec
Gutenberg (1913) depth 2900 km	
	OUTER CORE (FLUID) density = 9.7 gm/cc V_P = 10.3 km/sec
Lehman (1936) depth 5100 km	
	INNER CORE (SOLID) density = 16.0 gm/cc V_P = 11.2 km/sec
	EARTH CENTRE mean depth = 6370 km

Figure 2.5 Simplified model (zones and discontinuities) of the structure of the Earth interior

2.4.2 The Earth's Crust

The Crust of the Earth is the relatively thin layer represented by the outermost rock cover of the Earth. It is generally irregular as regards its thickness and inhomogeneous as regards its rock-composition. The crustal thickness is variable, ranging from 5 to 10 km under oceans and 40 to more than 70 km, under continents. The mean thickness of those parts of the crust under normal non-mountainous areas is around 30 km.

The Crust can be subdivided into two crystalline layers overlain by the outer heterogeneous rock cover which is made up of rocks which were subjected by various geological activities throughout the geological history. The two crystalline layers (called SiMa and SiAl) are separated by a discontinuity called the (Conrad Discontinuity). This discontinuity is observed only at certain parts of the continental crust. Anatomy of the Earth crust is presented in (Fig. 2.6).

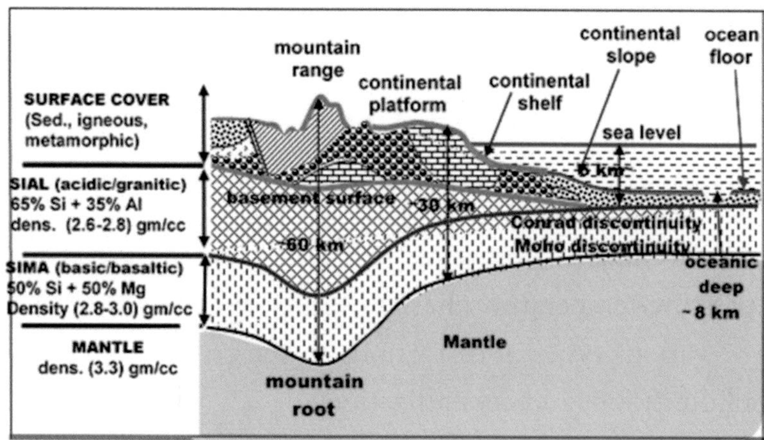

Figure 2.6 Anatomy of the Earth Crust

The crustal divisions are briefly defined as follows:

(i) The Sima Layer

This layer forms the basal part of the Crust, and it exists continuously under the continents as well as under the oceans. It consists of the two main chemical elements; Silicon (Si) and Magnesium (Mg) with the proportion of 50% for Silicon and 50% for Magnesium, hence it is given the name; Sima, indicating its composition. It is of basaltic nature having an average density of (2.9) gm/cc.

(ii) The Sial Layer

This layer forms the top part of the Crust, and it exists only under the continents. It consists of the two main chemical elements; Silicon and Aluminum with the proportion of 65% for Silicon and 35% for Aluminum; hence it is given the name Sial. It is of granitic types of rocks with density of about (2.7) gm/cc.

iii) The Surface Rock Cover

The outer surface cover of the Crust is formed from the oceans and continental masses which consist of various types of rocks overlying the more homogenous crystalline Sial layer found under the continents. In the geological literature, these rocks are considered as part of the Sial layer. The rock cover of the crust which consists of heterogeneous rock medium with various types of structural deformations may be more appropriate to be considered as a zone (the surface rock cover) separate from the Sial layer which is of distinctive homogeneous and crystalline nature.

2.4.3 The Earth Natural Forces

There are two groups of forces existing in nature which have been acting on the crustal rocks (especially on the crustal outer part), during the past long geological times. The two types of these forces are of internal and external origins.

- Forces of Internal Origin:

These are: the convection currents occurring in the Upper Mantle, large-scale continental tectonic movements, pressure-temperature changes, natural gravitational and magnetic field. All these phenomena cause various types of distortions to the crustal layers like folding, faulting, fracturing, melting, solidification, and crystallization.

- Forces of External Origin:

These are mostly changes of the weather-conditions, biological activities, and chemical processes. These natural activities are causing alterations of rocks such as denudation (weathering, erosion, and transportation), deposition, compaction, and cementation of rock fragments and particles.

Throughout geological history, and due to the effects of these forces, the crustal parts have suffered from compositional changes and re-distribution. Typically, rock layers are folded, faulted, moved about, and structurally deformed. As a result of these and other related factors, various types of rocks and minerals deposits were formed. The main types of rocks dominating the outer part of the Crust are sedimentary, igneous, and metamorphic rocks, which may be associated with water, oil, and mineral deposits.

2.5 Rocks

The exposed part of the Earth Crust is not permanently fixed in position and not permanently fixed in shape or in composition. The outcropping rocks are subject to denudation and other types of changes. Due to the effect of the active natural forces, broken parts of rocks and sediments are transported towards valleys and coastal areas. These sediments are then compacted by the overburden of incoming sediments forming the sedimentary rocks. Under certain pressure and temperature changes, these rocks are changed into various forms of rock-texture or to molten rocks (magma) which, on cooling, may change to crystallized igneous rocks. In consequence to the effects of the natural; forces, many different rock types are formed. Rock alterations take the form of a continuous cyclic process, the natural rock cycle.

2.5.1 The Natural Rock Cycle

The igneous and metamorphic rocks are changed into sedimentary rocks when they are exposed to denudation processes. Under heat and pressure conditions, sedimentary rocks may change into metamorphic rocks which in turn, may change into crystalline igneous rocks or to other sedimentary rocks according to the prevailing geological conditions. These interchanges of rocks from one type to another, is believed to be a non-stopping cyclic process (the natural rock cycle) taking place throughout the geological times. A simplified model for the rock cycle, is shown in (Fig. 2.7).

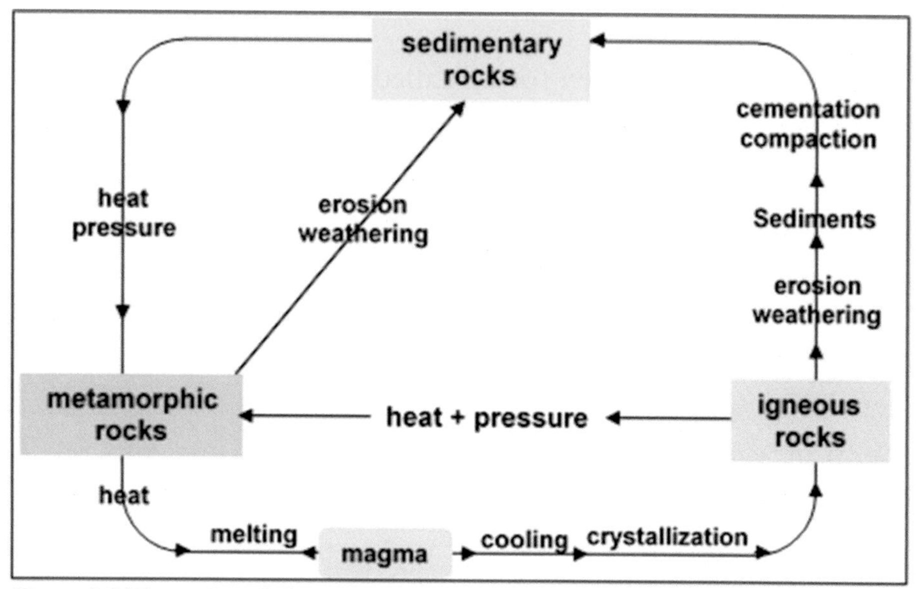

Figure 2.7 The rock cycle in nature

2.5.2 Types of Rocks

The enormously diversified crustal rocks can be grouped under three principal types. These are the igneous, sedimentary, and metamorphic rocks, defined as follows (Fig. 2.8):

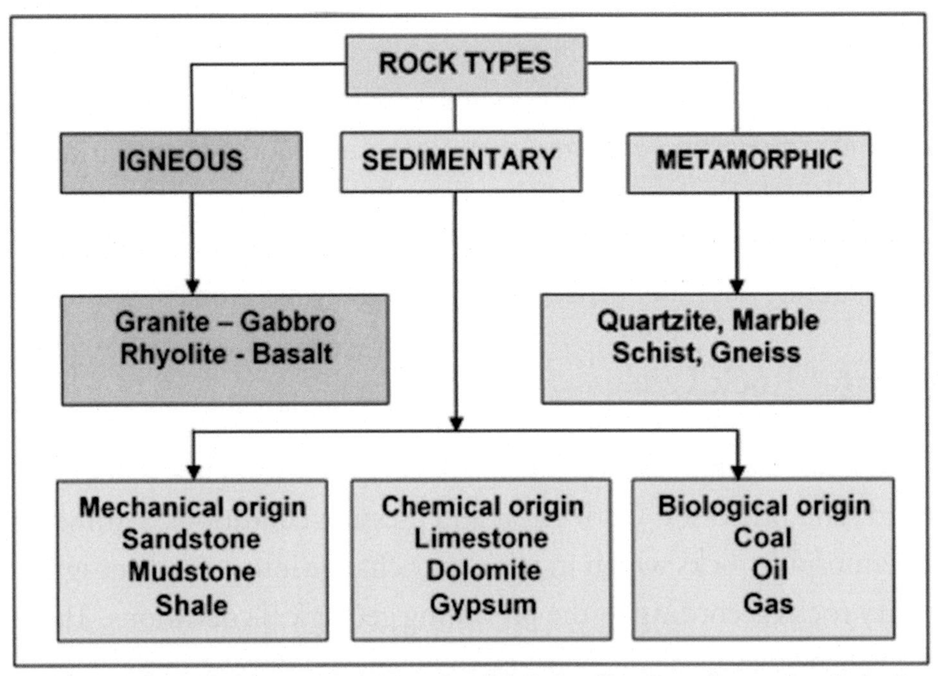

Figure 2.8 Block diagram showing a simplified classification scheme of rocks in the Earth Crust, with selected examples.

(i) Igneous Rocks

Igneous rocks are formed from molten rocks (called magma) or partially molten rocks that penetrate the Crust and get cooled before reaching the surface forming what are commonly called the (intrusive rocks), or after being exposed at the surface forming the extrusive rocks. The rock texture of the intrusive rocks is usually made up of coarse grains, in contrast to the extrusive rocks which are made up of fine grains.

An igneous rock is sometimes referred to, as acidic or basic rock depending on its mineral and silica percentages. An igneous rock, with high percentage of silica (as quartz), is called acidic, whereas those with little percentage of silica and rich in alkaline minerals (as plagioclase-series feldspar) are called basic igneous rocks. Typical examples of these two types of rocks are granitic rocks (acidic) and gabbroic rocks (basic) respectively.

Examples for acidic igneous rocks are granite (coarse grain) and Rhyolite (fine grain), and for basic igneous rocks, gabbro (coarse grain), and basalt (fine grain) rocks.

(ii) Sedimentary Rocks

Sedimentary rocks are formed from accumulations of bits and pieces resulting from disintegration of any previously existing type of rocks. The produced rock fragments are transported by natural forces to other locations where they are deposited under marine or non-marine environments. The accumulated sediments include plant and animal remains. With time, these sediments get pressed, cemented, and hardened to form thick stratified rocks, the sedimentary rock formations. Earth tectonic forces cause uplifting, folding, and/or faulting of the sedimentary layers bringing certain parts of these rocks to surface where they are exposed once more to weathering and denudation conditions. In fact, sedimentary rocks have their own cycle which is governed by weather changes, gravity, tectonic activities, and other Earth natural forces. Depending on the prevailing deposition environments, different types of sedimentary rocks are being formed. Examples of such types are: sandstone, shale, limestone, dolomite, and gypsum.

(iii) Metamorphic Rocks

Pressure, temperature, and chemical activities have the effect of changing rocks (igneous and sedimentary) into a new kind of rocks called (metamorphic rocks). Examples of metamorphic rocks created from sedimentary rocks are quartzite from sandstone, marble from limestone, slate from shale, gneiss or schist from older igneous rocks.

2.5.3 Sedimentary Deposition-Environments

Since oil deposits are mostly found associated with sedimentary rocks, additional useful information on sedimentary rocks shall be given here. A fundamental issue in this regards is the sediment deposition-environments and their created sedimentary rock-types.

The general model accepted for the way sedimentary rocks are made, is that sand particles and other fragments parted from older rocks, are transported down land slopes towards sea shores (or to other low-lands) by acts of natural forces (like gravity and wind). The prevailing deposition conditions under which the process is occurring is called marine environment when deposition takes place in the sea-water, and non-marine or land environment when deposition is in land areas. In general, the physical properties of a sedimentary rock change according to the sedimentation environment under which it has been formed.

The term (facies unit) is used for a sedimentary rock specified by its type of lithology and by its fossil-content. Thus, a rock, formed in shallow-water deposits (sandstone facies), differs from that deposited in deep water conditions (shale-facies). The most dominant depositional environments and the sorts of rocks formed are shown in (Fig. 2.9).

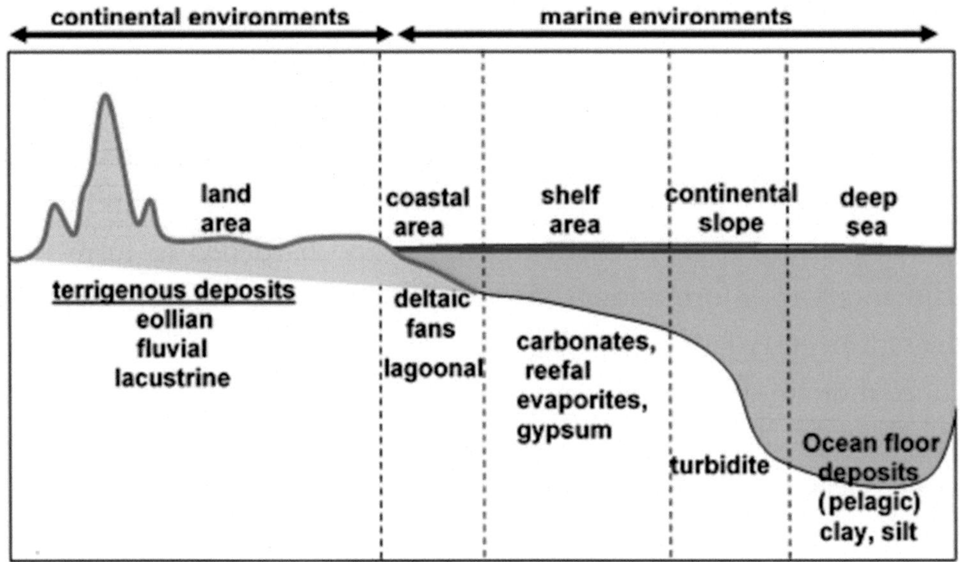

Figure 2.9 Deposition environments and their associated rock types.

Land (or terrigenous) deposits are derived from terrestrial environments. These sediments are carried by rivers, glaciers, or by wind finding their ways to low lands and sea shores. Some of these sediments are deposited on land areas (continental deposits) as those deposited in the flood plains (fluvial deposits), in lakes (lacustrine deposits), by glaciers (glacial deposits), or by wind (eolian deposits).

2.5.4 Sedimentary Rock-Types

Through rivers and tributaries, sediments are transported to river mouths where they lose energy and thus settle down forming what are known as deltaic deposits. In certain cases sediments get deposited on river-sides at times of floods forming the (fluvial deposits), or may find their ways to isolated water bodies as swamps, and lakes. The resulting rocks are termed (lacustrine deposits).

According to the particular type of deposition-environment, under which they were formed, sedimentary rocks can be divided into two main types of rocks. These are: the rocks of mechanical origins and rocks of chemical origins, as summarized in the following table.

Rocks of Mechanical Origin		
fine grain Argillaceous Rocks (clay)	medium grain Arenaceous Rocks (sand)	coarse grain Rudaceous Rocks (boulders, gravel)
claystone mudstone (compact) siltstone shale (laminated)	Sandstone quartzite (quartz rich) arkose (feldspar rich) greywacke (not sorted)	conglumerate breccia

Rocks of Chemical Origin		
Carbonates	Sulphates	Chlorides
Limestone (Ca CO_3) Dolomite (Mg CO_3) Marl (limestone+clay)	Anhydrite (Ca SO_4) Gypsum (Ca $SO_4.2H_2O$)	Halite (Na Cl)

2.5.5 The Sedimentary Rock Formation

The process of forming a sedimentary rock-layer passes through several phases before forming the final solid rock-bed which is commonly known as the (sedimentary rock-formation). The Earth natural fields (as gravitational and magnetic fields), in addition to weather, solar, and terrestrial heat changes, provide the main sources of energy responsible for the transformation processes which are leading to creation of the rock formation.

Creation of the sedimentary rock formation passes through a sequence of phases governed by changes of the prevailing environments. These changes are summarized as follows:

(i) Sorting of the sediments constituents during transporting phase. Coarse grain sediments are generally deposited in shallow water zones and the finer grain material is deposited in deep-water zones. Superposition of sediments in the deposition zones continues with time, leading to more and more sediments which are accumulated over previously deposited older material. The end result is a heap of sediments made up of conformable sedimentary layers.

(ii) Lithification of the sediments caused by the cementation, compaction, and dehydration, producing the familiarly known hard sedimentary rock-beds.

(iii) Deformation of the formed sedimentary rock-layers by mechanical factors (as tectonic forces), causing folding, fracturing, and faulting. In addition to these structural changes, possible diagenesis processes may occur. Changes in rock texture and mineral contents may take place as a result of diagenesis. A process in which rock nature (physical and chemical properties) may be changed in relatively low pressure and low-temperature environments.

. The development phases involved in these processes can be summarized as follows (Fig. 2.10):

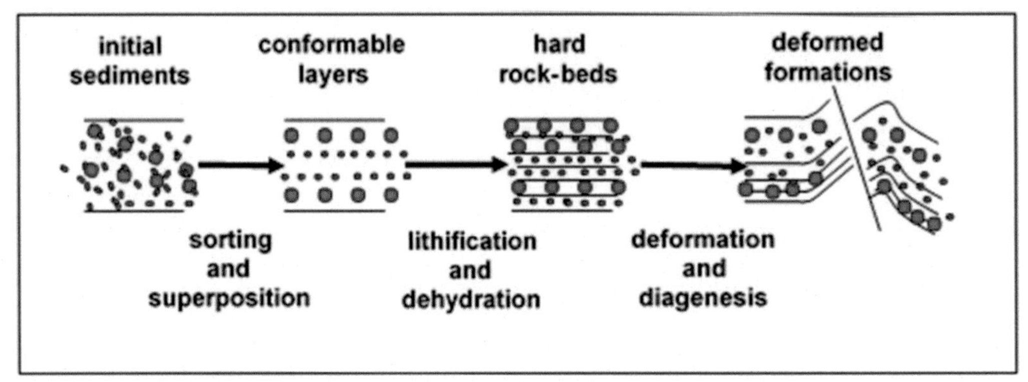

Figure 2.10 Development phases in forming of the sedimentary formation

2.6 The Geological Time Scale

The age of the Planet Earth is estimated to be about five billion years. Palaeontological evidences indicated that biological life started 600 million years ago. Based on breaks in the development of major geological changes, the Earth geological history was divided into three major intervals, called (Eras). These are: Palaeozoic, Mesozoic, and Cenozoic eras. Each of these eras is further subdivided into sub-divisions called (periods) and the rocks, deposited within each period, are named rock-systems. The Cenozoic era, for example, is subdivided into Tertiary, and Quaternary periods. These divisions are summarized in the following table (Fig. 2.11).

ERA	PERIOD	EPOCH	
Cenozoic	Quaternary	Holocene (recent) Pleistocene (glacial)	
	Tertiary	Neogene	Pliocene Miocene
		Paleogene	Oligocene Eocene Paleocene
Mesozoic	Cretaceous Jurassic Triassic		
Paleozoic	Permian Carboniferous Devonian Silurian Ordovician Cambrian		
Precambrian	Proterozoic Archaeozoic Eozoic		

Figure 2.11 The geological time scale based on (Holmes, 1975, p156)

A complete geologic time scale compiled by the Geological Society of America is available at (www.geosociety.org/science/timescl.htm).

It is useful to note here that during the geologic time, several mountain-building geological episodes (orogenic movements) have taken place. Geological research led to identification of three major orogenies. These are: the Caledonian orogeny which occurred during the Palaeozoic era, the Hercynian (Variscan) orogeny by late Palaeozoic era (Carboniferous-Permian), and the Alpine orogeny which seems to have started in early Tertiary, when its major phases of mountain building occurred in the Paleocene-Eocene periods and continued up to present times.

From the palaeontological studies (studies of the preserved fossils), animal and plant developments, during the geological system-sequence, have been established.

2.7 Structural Geology

The outer rock cover of the Earth Crust is characterized by being heterogeneous as regards composition and of non-uniform geological structures. Rock layers are deformed both in large and small scales, and they may include oil, water, and great varieties of other mineral deposits, in solid, liquid, and gaseous states.

Structural Geology is the study of deformation distribution (strain field) and the forces (stress field) that caused the realized rock deformations. Large-scale structural deformation (as mountain building, continental rifting, and creation of geosynclines) are commonly referred to as tectonic activities (or tectonics).

2.7.1 Crustal Tectonics and Continental Drift

The crustal large-scale deformations are believed to be related to the convection currents occurring in the Earth Upper Mantle which is found immediately beneath the base of the Earth Crust. Convection currents are believed to be behind creation of the crustal major tectonic features like the continental masses, continental shields, mountain ranges, ocean troughs, and large sedimentary basins.

The phenomenon of large-scale continental movements was based on a theory (continental-drift theory) developed in the early twentieth century, especially by the pioneering work of the German geologist, Alfred Wegener in 1915. According to this theory, Wegener proposed that the presently known continents were originally one land block then started to separate and drift away from each other until they reached the presently known positions. This proposal was based on the following geological observations:

- The western coastal lines of Africa are similar to the eastern coastal lines of South America.

- Types of rocks and fossils of the Cambrian rocks of NW Scotland are similar to those in Eastern Canada.

- Types of the Permian rocks of South Africa and India are similar to those found in Antarctca.

- Palaeomagnetic measurements indicated that the magnetic pole has moved over a path supporting that the two continents North America and Europe have drifted away from each other since early geological times.

2.7.2 Plate Tectonics

The Earth lithosphere, which includes the Crust and the Upper Mantle, is broken into continental blocks called tectonic plates. Formation and movements of the continental blocks is thought to be activated by the convection currents taking place in the Upper Mantle. The continents, considered as independent tectonic plates, have moved with respect to each other during the geological ages. Two main types of boundaries resulted between any pair of such plates; convergent (collision) zone and divergent (spreading) zone. The so-created boundary-zones are commonly known to be hosts of earthquakes, volcanoes, mountain building, and oceanic trenches. The global activity, which resulted in creating the continental blocks, is commonly referred to, as (plate tectonics).

The continental blocks (tectonic plates) include the continental as well as the oceanic lithosphere. The crustal part of the continental plates is much thicker (several tens of kilometers) than the oceanic plates (several kilometers). The hard-nature of the lithosphere compared with underlying Mantle-material (the asthenosphere), helps in the movements of the tectonic plates.

2.7.3 Crustal Tectonic Features

The formation of the crustal large-scale tectonic features is thought to be caused by the effect of the convection currents moving in the Upper Mantle. A plate tectonic theory is developed to describe the mechanism of the crust main tectonic features. According to this theory, it is proposed that there are several loops of convection currents, active in the Upper Mantle which creates two main types of deformation-zones in the Earth Crust; crustal tension- and crustal compression-zones (Fig. 2.12). .

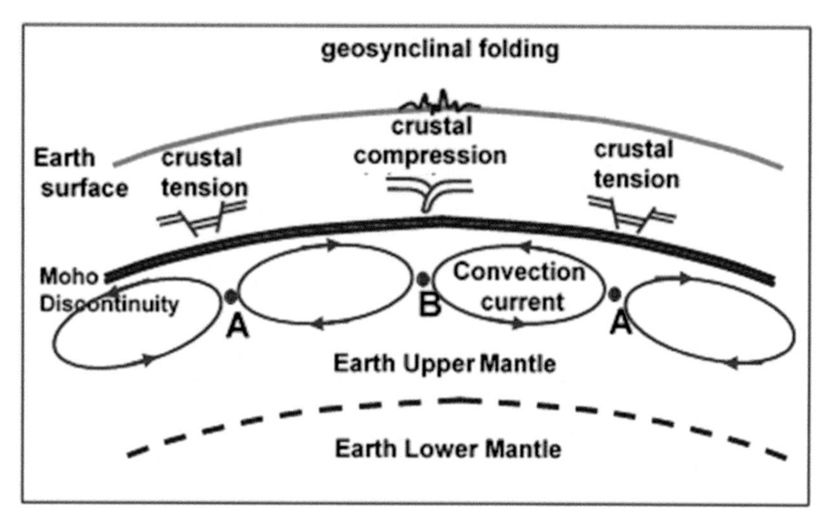

Figure 2.12 Sketch showing a simplified model for the causes of crustal tectonic features. Points A and B represent crustal tension and crustal compression respectively.

The two types of plate-boundary zones are:

(i) Crustal Tension Zone:
A crustal part located above the meeting point of two neighboring loops of convection currents whereby Mantle material are moving upwards. At this location the Earth Crust becomes under tension which leads to spreading effect causing parting of crustal plates. Examples of tectonic features created in such zones are sea-floor spreading with mid-ocean ridges (as the Mid-Atlantic Ridge) and continental rifting (as the Great African Rift Fault).

(ii) Crustal Compression Zone:
A crustal part located above the meeting point of two neighboring loops of convection currents whereby Mantle material are moving downwards. At this location the Earth Crust becomes under compression which leads to subduction effect causing collision of crustal plates. Examples of tectonic features created in such zones are geosynclinal folding and creating mountain ranges (as the Andes Mountains) and forming volcanic island arcs (as the Aleutian Islands and the Japanese island arcs).

At zones where two tectonic plates are passing each other, transform faulting is created. The Great San Andreas fault is a typical example of such a crustal tectonic feature.

2.8 The Crust Structural Features

The phrase (structural features) is usually applied to refer to the relatively small scale structural changes, compared with the large scale crustal tectonic changes as continental movements, mountain building, and ocean trenches. As we have just mentioned, tectonic forces are provided by the convection currents assumed to be working in the Upper Mantle which is believed to be made up of nearly molten rocks. These molten materials are moved by convection currents in the form of loops passing just below the base of the Crust. The end-effect of the convection currents is creating crustal stresses which cause large scale structural deformations of the lithosphere including its crustal part. The created deformations in the Crust take the form of regional tectonic features like mountain ranges and great continental rifting.

The smaller-scale deformations due to tectonic, and non-tectonic forces, lead to the relatively smaller structural features such as folding and faulting. In addition to these structural features, another type of changes occurs. These are the stratigraphic changes which include lithological, depositional, and denudation features). Both of these features (structural and stratigraphic) are strongly associated with formation, migration, and accumulation of oil, water, and other mineral deposits.

The main structural features which are commonly observed in the crustal geological layers are folding and faulting.

2.8.1 Folding of Rock layers

Due to regional or local stresses, crustal rocks undergo physical distortions like compression, extension, and shearing. Under certain conditions, rocks are bent (folded) rather than being broken. Rock folding may be brought about by the effects of regional tectonic or non-tectonic forces. As mentioned above, tectonic forces are normally having large-scale distortion effects which are leading to folding of regional extents, as geosynclines, geanticlines, and large scale monoclines which are commonly referred to as "regional dips".

The more localized types of folding are caused by differential compaction created by non-tectonic effects like those due to gravity, to chemical changes, glaciations, and growth of reefal deposits, salt diapers, and igneous intrusion. All these phenomena can lead to folding of the rocks. Folds, observed in nature, range in dimensions from few centimeters to many kilometers in extent. A folded layer in the form of a curve which is convex upward is called an anticline, and that concave upwards is called a syncline. The so-formed anticlines and synclines may be symmetrical or asymmetrical, or even over-turned. Further, these may be repeatedly folded, recumbent, anticlinorium, and synclinorium, with horizontal or inclined (plunging) axes.

In summary, structural features (mainly folds and faults) are caused by either tectonic forces or by non-tectonic forces. Under the first type (tectonic features), the outcome may be due to strong mountain building and thrust faulting forces (called orogenic forces), or weak simple folding and faulting (called epirogenic forces). Under the non-tectonic type of forces (as in differential compaction forces), folding and faulting can occur over growth structures (such as salt diapers or igneous plugs) or over faults (Fig. 2.13).

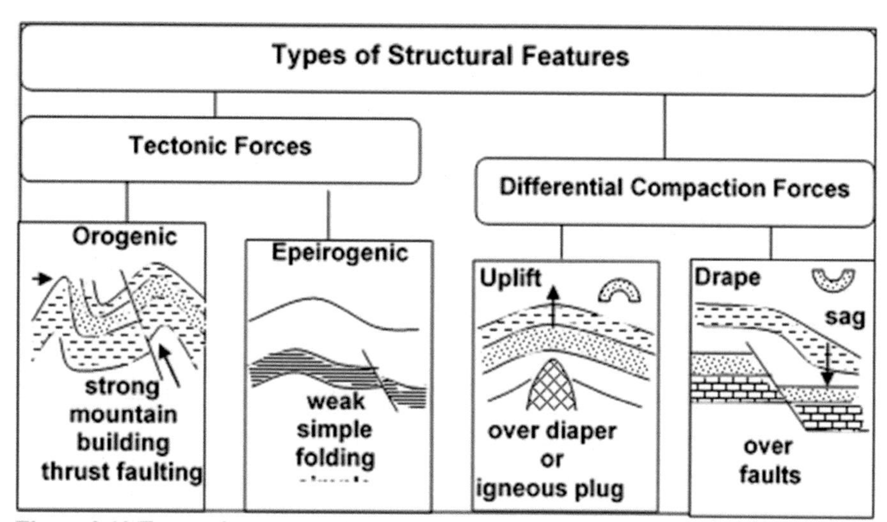

Figure 2.13 Types of structural features and their causing forces

2.8.2 Faulting of Rock layers

Rock layers under stresses exceeding rocks breaking-points, result in rupture. The process of rock rupture where-by the broken parts are shifted in respect to each other is called faulting. Faults are of several types depending on the geometry of the affected parts of the faulted layer. Examples of faults are:

- normal fault (also called gravity fault), caused by tensional forces.
- reverse fault, (or thrust fault) caused by compression force.
- wrench (or tear) fault.
- strike-slip fault
- rotational fault
- transform fault
- multiple faults as horsts and grabens.

The most commonly known examples of faults are presented in (Fig.2.14).

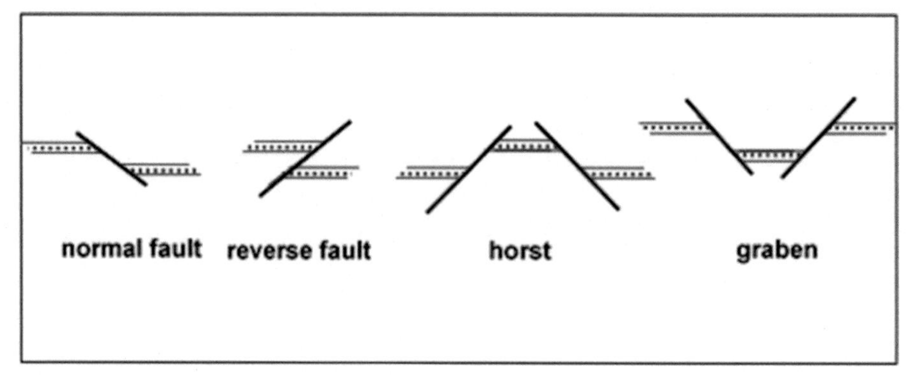

Figure 2.14 Common types of faulting

There is another type of faults which have some bearing to oil-reservoir studies. These are the (growth faults). A growth fault is believed to have been formed during the sedimentation process, especially in deltaic environments, or during vertical movement of the crystalline basement on which the sedimentation process was taking place. Normally, it is observed that both of the fault throw and the bed thickness in the down-thrown side of the fault, are increasing as depth increases. Furthermore, the dip of the fault plane is not constant, as in the case of the other types of faults, but it is decreasing with depth (Fig. 2.15).

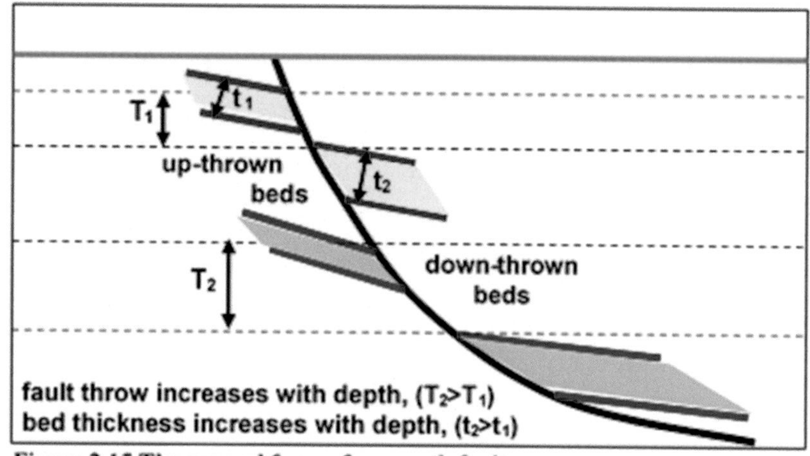

Figure 2.15 The general form of a growth fault

Properties of growth faults, that have relation to oil-reservoir studies, are the changes in porosity and permeability of the beds across the fault. It is noted that down-thrown beds are of lower porosity and permeability than that of the up-thrown beds. Full treatment of the growth faults is given by (Chapman, 1976, p. 83-93).

2.8.3 Mountain Ranges

One of the phases of the crustal changes is formation of the mountain ranges. Over the long geological time, the sediments infill deposited in the major ocean troughs can reach several kilometers in thickness. When these accumulated sediments are subjected to compression forces, (as it happens in subduction zones of the Crust), structural uplifts (and depressions) are caused in the affected rocks. The so-created uplifted parts represent the mountain ranges.

The mountain-building phenomenon (called the orogenic process), is part of a natural cycle that happens to the crustal layer. This cycle is represented by the sequence:

- Uplift of part of the Crust
- Denudation of the uplifted rocks
- Redepostion in the oceanic troughs
- Reduction of the mountain material and increase of deposited sediments
- Balance restoration by sinking of deposited sediments and uplifting of original sediment-source, creating the mountain ranges.

2.8.4 Rift Zones

Alongside with the creation of the mountain ranges, natural forces create, in certain parts of the Crust, faults of various types and sizes. One of these features is a type of large-scale faulting which is called (rift faulting). This large-scale feature is created due to extensional crustal movements or due to major strike-slip faulting that leads to pulling apart of crustal blocks. When the zone between two normal faults (as in graben structures) slides down due to faulting, a valley (called rift valley) is formed. Affected basement rocks, found beneath sedimentary basins which are experiencing extensional strain, are creating basin structural changes favorable for oil entrapment. An outstanding example of this phenomenon (shear movement) is the Great African Rift Fault which resulted in a surface feature known as the Central African Shear Zone (CASZ) crossing the whole of Africa from the Cameron in the West to Sudan in the East (Fig. 2.16).

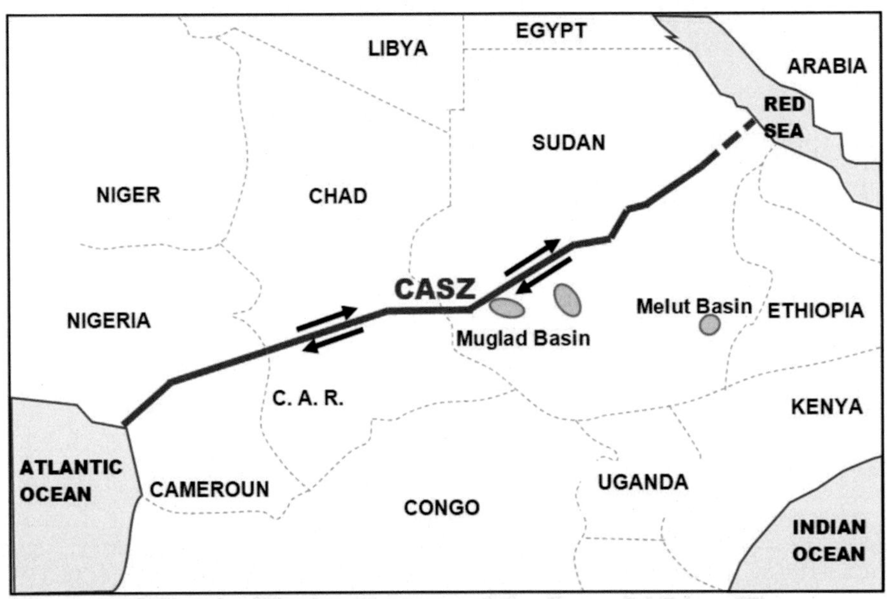

Figure 2.16 Sketch of the location map of the Central African Shear Zone (CASZ), based on (Fairhead, 1988).

This large-scale crustal tectonic feature (CASZ) has been active during Jurassic-to- Miocene and was the main factor that controlled the geological setup of that African region. Further, it has governed the types and sizes of sedimentary basins within that zone. Those created basins contain numerous oil fields.

2.9 The Crust Stratigraphic Features

Not only changing of rock-beds geometrical form (folding and faulting), the natural Earth forces are also incurring changes on the rock lithology, chemical composition, and other physical properties such as rock density, porosity, permeability, and mineral contents. All these features (lithplogy, chemical and physical properties) are collectively represented by the crustal stratigraphic features. Mineral and hydrocarbon contents are considered as part of the stratigraphic features of the outer parts of the Earth Crust.

Geological studies have shown that stratigraphic features of sedimentary rocks are governed by the sedimentary environments that were prevailing during the precipitation process. Typical types of sedimentary deposits are non-marine (or land) deposits, river fluvial deposits, high energy and low energy shelf deposits, and deep water argillaceous deposits. Types of deposits are dependent upon the particular deposition environment, as shown in (Fig 2.17).

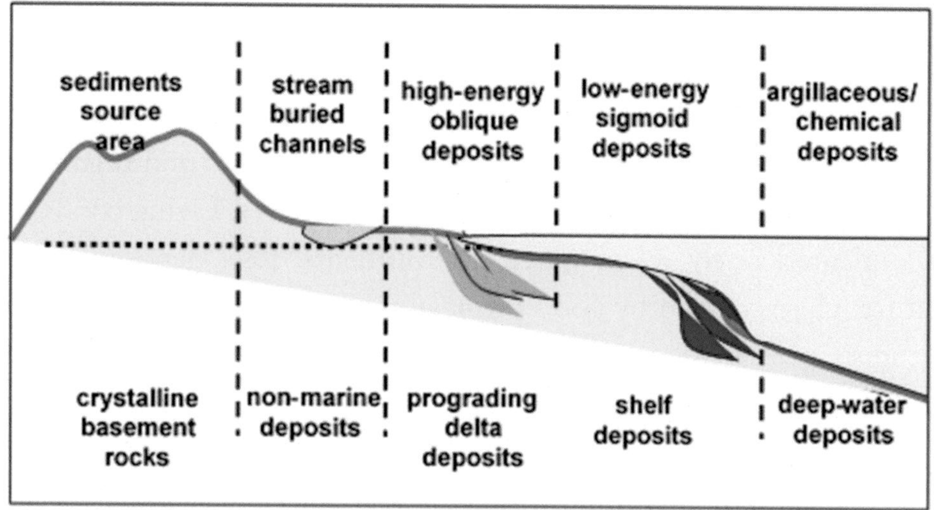

Figure 2.17 Dependence of sedimentary rock-types upon the types of deposition environments.

Stratigraphic features which have important roles in forming hydrocarbon reservoirs are: sand lenses, reef deposits, salt domes, and unconformities. Sketches of common examples of these features are presented here-below (Fig. 2.18):

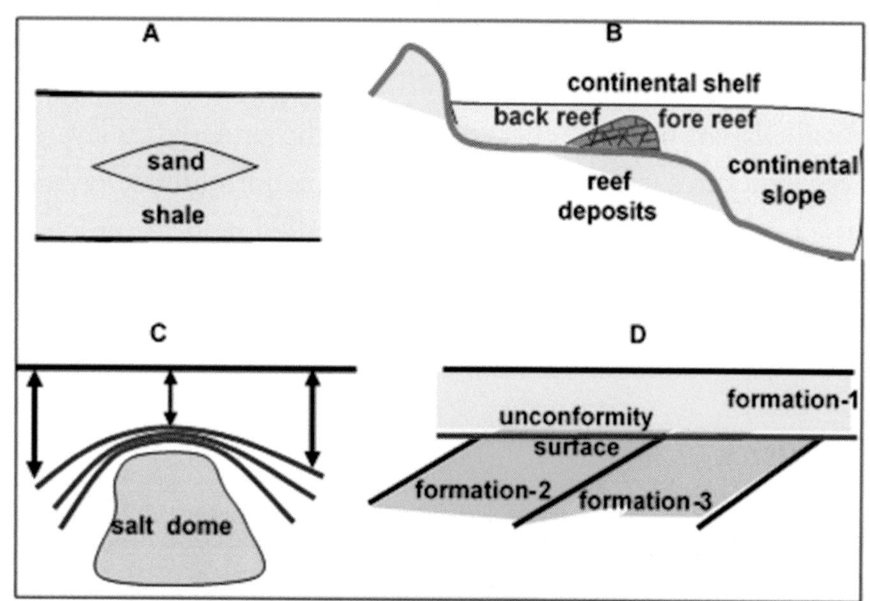

Figure 2.18 Common types of crustal stratigraphic features: (A) sand lens, (B) reef deposits, (C) salt dome, (D) unconformity.

2.9.1 Sand Lenses

Sand deposits filling confined troughs, or channels, or those forming marine sand-bars near, and parallel to, sea coastal lines are forms of isolated sand bodies which are generally referred to as sand lenses. These stratigraphic features form sandstone bodies suitable for hydrocarbon accumulation.

2.9.2 Reef Deposits

These deposits are mounds formed mainly of skeletal remains of coral organisms which live in special marine environments, normally on continental shelves. Living corals need relatively shallow, clear, and moderately warm water environments. Reef deposits are of carbonatic (limestone) nature, characterized by good primary porosity which, due to certain chemical changes (diagenessis processes), secondary porosity may be developed. The good porosity property makes reef bodies, favorable media for being suitable hydrocarbon reservoirs.

A typical reef body is made up of porous calcareous material found within shale or mudstone deposits

2.9.3 Salt Domes and Rock-Flow Structures

Due to overburden pressure, some rock-types (such as salt and clay) deform by plastic-flow process and migrate vertically and horizontally, forming intrusions within other types of sediments. Under such conditions, salt deposits, which act as plastic materials, are pushed to migrate away from their original positions. The migrating salt, called diapers, may push the overlying strata causing structural deformation like anticlines, synclines, domes, and faults. These effects are called (salt-tectonic effects) whereas the underlying layers are left flat and undisturbed. These structures (resulting from diapers) are normally described as halo-kinetic, but when the cause is compressive tectonic force, it is called halo-tectonic.

With diaperism, both folding and faulting may occur. Salt structures (salt domes, plugs, and pillows) can create traps for oil-accumulation.

2.9.4 Unconformities

When sedimentary rock layers are folded and tilted, then eroded away down to a flat surface, may get covered by new sediments forming the unconformity feature. Strictly speaking, the term unconformity (or angular unconformity) is used to denote the surface separating the older structurally-distorted beds from the newer overlying beds. When the unconformity surface is at an angle with the older deposited beds, it is called (angular unconformity), and when all involved beds are parallel, the surface represents lapse of time with no sedimentation (quiescent period) it is called (disconformities).

2.9.5 Facies Changes

Changes in lithology of sedimentary formations are caused by changes in the depositional environments. Sea-level changes (rising and falling) are the main factor controlling the lithological types of resulting sedimentary layers. Within a basin, sedimentary facies which are of coarse deposits are found near basin margins. The rock sedimentary facies are normally found to be changing into more fine (silt or shale) sediments near basin centre. Studies of sedimentary facies help in the indication of the position of the sediments source area.

2.10 Sedimentary Basins

The sedimentary basin is a large-scale body of sedimentary-rocks formed as a result of sediments deposition and the tectonic activities (folding and faulting) which have occurred during and after the deposition process. One of the effects of the convection currents of the Upper Mantle is formation of great geosynclinal and large-scale fault systems. The so-formed great depressions are filled with sediments forming large (regional) sedimentary basins. Such large sedimentary basins are also formed from influx of sediments in the river deltas (deltaic deposits). Thickness of the sedimentary infill of the basin may reach (10-15) km.

Depth of the crystalline basement rocks, bellow the sedimentary basin, varies depending on the measurement location. Normally, it is deeper near the basin centre and shallower near basin margins. At certain cases, the basement rocks are seen to be outcropping on the rim of the basin.

Based on the way of creating the basin-depression, three types of sedimentary basins can be identified. These are:

(i) Crustal Compression Basins:
These are formed in the subduction zones created at some of the plate margins. Due to the compression forces which are created in these zones, the beds in the created sedimentary basin are folded forming a type of basins called geosynclinals basins which include structural traps, favorable for hydrocarbon accumulations. Most of Middle East oil fields are in this type of sedimentary basins.

(ii) Crustal Tension Basins:
These are formed in the crustal-tension zones, where sea-floor spreading takes place with the result of great rifting and various types of associated faulting. The Great African Rift Zone is a typical example of this type of basins.

(iii) Deltaic Deposition Basins:
These are formed as a result of purely stratigraphic activities. The sediments are transported by rivers down to the seas where they are deposited in the continental margins. The accumulated sediments cause the sea floor to get depressed leading to accommodation of more incoming sediments. In this way thick sedimentary basins (deltaic type) are formed. Stratigraphic hydrocarbon-traps are common in this type of basins.

Based on the form of symmetry of basins geometrical shapes, four common types can be identified. These are: symmetrical basin, asymmetrical basin, graben basin (bounded by two normal faults), and half-graben basin (bounded by one normal fault). These types are shown in (Fig. 2.19).

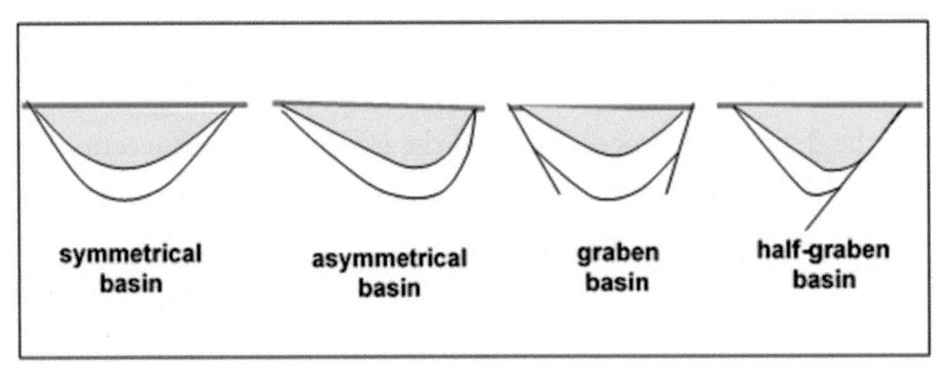

Figure 2.19 Classification of sedimentary basin on basis of geometrical shape

2.11 Continental Shelves

Continental margins covered with ocean water at depth of (100-200) meters, are known as the continental shelves. Water depth at shelf areas gradually increases in the direction of the ocean deeps. Crustal thickness beneath continental masses decreases from about 50-60 km under continents, to about 5-8 km under the oceanic deeps. Geological knowledge indicated that the crystalline Sial layer of the Earth Crust is absent in those crustal parts beneath the oceans.

From geological point of view, continental shelves are considered to be integral parts of continents, having same geological nature (structural and stratigraphic features) as the exposed continents.

Marine oil-exploration activities in shelve areas proved to have hydrocarbon potentialities comparable with those found in land areas. Typical discovered continental shelf oil-fields are those found in the Gulf of Mexico to the south of the United States and in the North Sea to the West of Norway.

Chapter-3

3. OIL GENERATION AND MIGRATION

Petroleum geologists have agreed on a model for the petroleum occurrence. In essence, this model is represented by a three-stage process which is: generation, migration, and accumulation. In the geological literature, one finds that the question of mechanism of how each of these processes has been accomplished is still debatable. In this chapter, the more agreed-upon theories of petroleum origin and migration shall be briefly presented.

3.1 Development Phases of Oil

Petroleum geologists believe that an oil body has been formed through three main phases of development. These are generation, migration, and accumulation as shown schematically in (Fig. 3.1).

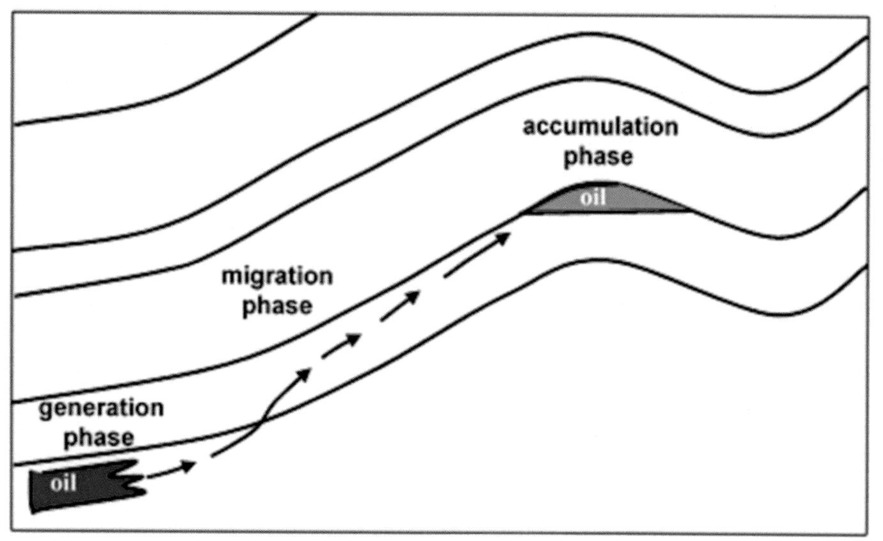

Figure 3.1 Sketch showing the three development phases (generation, migration, and accumulation) of an oil body.

These three processes are briefly defined here-below:

3.1.1 The Generation Phase

Oil is generated alongside with the precipitation process of a fine-grained material of rocks of argillaceous or calcareous nature. These rocks are containing organic material which is later transformed into a homogeneous matter of complex chemical composition called Kerogen. This process is taking place with the help of a certain type of Bacteria that acts under anaerobic environments and under favorable pressure and temperature conditions. At a later stage, the Kerogen starts to change into hydrocarbon matter of quantity and quality governed by the prevailing pressure and temperature. The rock medium, within which the hydrocarbon matter is formed, is called the (source rocks) or (mother rocks) and the place, where the transformation process occurs, is normally referred to as the (oil-kitchen).

3.1.2 The Migration Phase:

After the generation phase, and due to the effective overburden pressure, the generated oil migrates from the mother rocks to the porous neighboring rock media. This motion (called primary migration), is followed by a second migration process, where the oil continues in motion, under favorable conditions of porous and permeable pathways, to the final accumulation zones. The oil quantity and migration directions are governed by the properties of the rock medium, through which the oil is moving, as well as on the physical properties of the migrating oil itself.

3.1.3 The Accumulation Phase:

The migrating oil gets accumulated in a zone within the rock-medium when its motion is stopped and prevented from further movement. A closed zone, in which oil is trapped, is called an (oil trap). The fundamental condition of an oil trap is that it is made up of porous and permeable rocks which allow the incoming oil or gas to fill in the pores of that medium. Further migration of the accumulated oil is prevented by geological barriers such as impervious cap rocks or a sealing barrier caused by a fault or other such-like mechanism.

3.2 Theories of Oil Origin

The origin of oil is one of the problematic issues which were subject to speculation and laboratory experimentation since discovering oil in large-scale commercial quantities, around the mid nineteenth century. There are two main groups of theories developed for the origin of oil. Some theories (abiogenic theories) suggest that oil is of inorganic origin, whereas others (biogenic theories), suggest that it is of organic origin. Evidences observed by geologists are more in support of the organic theories as they lend themselves to be more realistic and more readily-accepted principles which explain the mechanism of oil generation and its subsequent migration. In general, the two groups of theories are: the inorganic theories states that oil is generated from inorganic material and the other states that it is generated from organic material.

3.2.1 The Inorganic Theories

There are several theories (or hypotheses) which were presented in support of the inorganic origin of oil. In brief these are the following:

(i) The Chemical Theory

This theory supposes that the hydrocarbon compound was formed from a chemical reaction of water and metallic carbide or from reaction of water with carbon dioxide or with chemically-basic compounds. Based on laboratory experiments, it was found that a metal carbide (like iron carbide, FeC) when treated with water vapor under high temperature and pressure, Ethane compound ($C_2\ C_6$) can be obtained. Ethane is one of the Paraffin series, which is one of the principal constituents of hydrocarbons.

This theory requires that the involved chemical reaction must have been taken place at large depths where such high temperature and pressure shall be prevailing. However at such great depths host rocks are not expected to have the required porosity to host the chemically-produced oil with the huge quantities as found in the discovered oil-fields.

(ii) The Volcanic Theory

According to this theory oil is considered to be one of the products of volcanic activities because of presence of hydrocarbon gases with the expelled volcanic gases. It is assumed that the observation is indicating presence of hydrocarbon matters within the rocks from which the molten magma was produced. The theory supposes that these hydrocarbon gases are later converted into oil under suitable chemical conditions.

Like lava-volcanoes, mud volcanoes have also been reported to have associated hydrocarbon gases. In both types of volcanoes, it is assumed that the hydrocarbon matter has been formed first within sedimentary rocks then they were transformed to the volcano's region. The theory is in need of explaining of the abundance of oil in large quantities in sedimentary rocks and not in igneous rocks.

(iii) The Cosmic Theory

This theory postulates that primeval hydrocarbon gases were incorporated in the solar planets (including the Earth) during the formation process. The hydrocarbon gases were then condensed into oil liquids which are collected in the crustal rocks. If this model of the generation process were the case, then it is expected to have uniform distribution of oil within the crustal rocks and not in different isolated locations as it is found in the case of the planet Earth.

3.2.2 The Organic Theories

The organic (or biogenic) theory is the more accepted theory for the way oil matter has been formed. According to this theory, oil is of organic origin, meaning that it was formed from biological material (plants and animals). According to this theory, oil has been generated through a sequence of events starting with the deposition of a mixture of sediments and the biological remains (micro-organisms) into the sea water.

With further accumulation of sediments these remains are buried under great amount of sediments. Under high pressure and temperature, and with the action of special types of anaerobic bacteria, the biological matter is converted into hydrocarbon compounds, in liquid and gaseous states. The conversion process must have taken place in the absence of oxygen.

The model now accepted for the mechanism of oil-generation process is that oil is formed from the remains of prehistoric marine animals and plants which have been settled to sea bottoms in huge quantities under anaerobic environments. Over long geological times, the deposited organic matter, mixed with argillaceous sediments, is buried under thick accumulations of sediments. Such kind of deposition environments brings about high pressure and temperature which cause chemical changes in the deposited organic matter producing a primitive hydrocarbon, called Kerogen. With further increase of temperature and pressure, Kerogen is converted into liquid and gaseous hydrocarbons. The generated oil is then distributed in the host sediments and under pressure it is forced to move through the pores of the overlying rocks until it is trapped or find its way to surface.

The most important supporting evidences for the biogenesis origin (oil derived from sea microbiological organisms) are the huge quantities of oil which are found in association with sedimentary rocks containing fossil micro-organisms. Further, the chemical compounds of oil constituents are the same as those of the biological bodies (plants and animals). In addition to that, it is observed that the organic remains found in the oil source-rocks are of micro-biological organisms, as foraminifera, algae, and pollens. Other supporting evidence to validity of this theory came from geochemical studies. Gas chromatography has indicated that the hydrocarbon molecules of the organic matter, which are found, scattered in sedimentary rocks (bituminous matter) has the same chemical composition as those of the crude oil.

In certain rare cases, oil was found in non-sedimentary types of rocks, as in igneous and metamorphic rocks. In such cases it is likely that the oil has been originally formed in some neighboring rocks of sedimentary nature, and moved to the fractured parts of the igneous (or metamorphic) rocks later on. It should be noted here that some recent studies postulated that oil found in non sedimentary environments could have been formed by inorganic (abiogenic) types of processes.

Genesis of oil based on the organic theory, is the most accepted mechanism for the oil generation process. It is hard to believe, as petroleum geologists say, that oil with this huge reserves, found associated with sedimentary rocks at limited depth ranges, can be formed by strict chemical reactions (i.e inorganic means).

3.2.3 The Kerogen

According to the organic theory (explained in the previous paragraph), oil has been formed from organic matter (biological remains) which were deposited alongside with the deposition of the sedimentary rocks. The deposited organic matter were then buried under thick covering sediments which are creating high pressure and temperature. Under these environments and with the action of a special type of anaerobic bacteria, that act under chemically-reducing conditions (in absence of Oxygen), and over long geological time, hydrocarbon matterl is formed. This generation process may be summarized as follows:

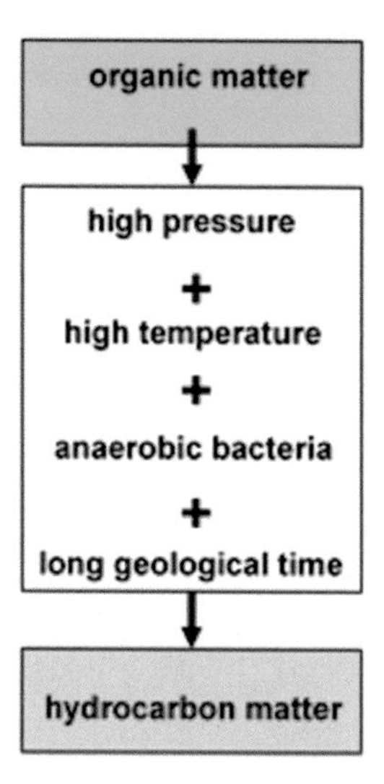

The rocks, with which oil is generated, are normally fine-grain, argillaceous sedimentary rocks (as clay, shale, and marl) which have been deposited under calm marine environments.

Being the medium within which oil has been formed, these rocks are called (source rocks) and the generation place is often referred to as the (oil kitchen). Petroleum geologists agreed that the transformation process (change of raw biological remains to hydrocarbon matter) has taken place via a transitional, or intermediary, stage represented by forming a homogeneous organic matter called (kerogen) which consists of a chemically complex material, insoluble in the known hydrocarbon solvents.

The role of Kerogen in the oil-generation process can be considered as an intermediary substance created from the deposited organic matter, then it (the Kerogen) is transformed into hydrocarbon compounds. The central position of Kerogen in the sequence of oil generation processes can be represented as follows:

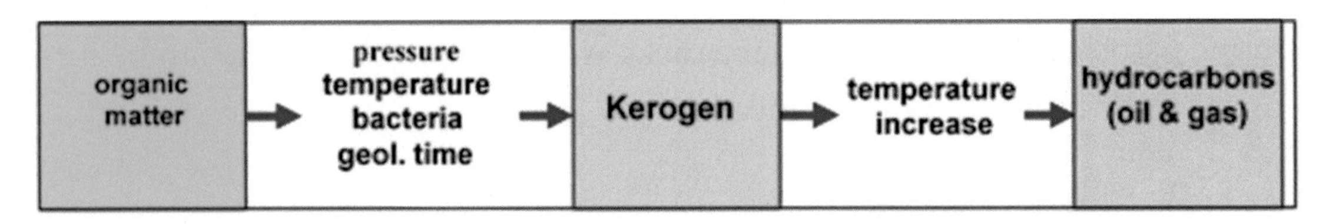

3.2.4 Types of Kerogen

Based on the chemical composition, three types of Kerogen are recognized (Selley, 1983, p. 9-10). These are:

<u>Type-1 kerogen (Algal)</u>

It is rich in algae and it tends to generate oil, and found in many oil shale rocks.

<u>Type-2 kerogen (Liptinic)</u>

It is rich in aliphatic compounds (hydrocarbons of straight, branched, saturated, or unsaturated hydrocarbon series) and low in ring aromatic compounds. This type of Kerogen can generate both of oil and gas.

<u>Type-3 kerogen (Humic)</u>

It is rich in aromatic and low in aliphatic compounds. This type is produced from wood remains and it is more of bituminous coal, lignite and peat. It is capable of producing gas.

It is stated (Selley, 1983, p. 10) that both of the two kerogen types (type-1 and type-2) tend to occur in marine environments, while type-3 kerogen tends to occur in continental (fluvial and deltaic) environments. This is in support of the generalization made by geologists that marine source rocks tend to generate liquid hydrocarbon (oil) but the continental rocks tend to produce gaseous hydrocarbon (gas).

3.2.5 The Three Phases of Alteration

After being buried, the organic matter undergoes three stages of alteration, summarized as follows (Stoneley, 1995, p. 33):

(i) Diagenesis Stage:
Diagenesis is defined to be an alteration process, which is mainly chemical, that occurs to the deposited organic matter under relatively low-pressure and low-temperature environments. At this stage, the organic deposits are subjected to chemical changes and to bacterial decay leading to transformation into the homogeneous matter, the Kerogen. The diagenesis process also results in sediments compaction and hence in reduction of porosity.

(ii) Catagenesis Stage:
The term catagenesis is used by petroleum geologists to mean a process that involves cracking of the organic kerogen, converting it into oil and hydrocarbon gases. The process of oil generation starts as temperature reaches 150^0 F. At greater depths, as temperature gets higher, oil generated from the Kerogen reaches a maximum level, then with further increase of depth the generated oil drops off to a minimum level. The range of depths over which oil is generated (called the oil window) is roughly (3-5) km. Within the deeper parts of the oil window the generated hydrocarbon consists of wet gas and liquid oil.

(iii) Metagenesis Stage:
At depths below the oil window where temperature becomes greater than 350^0 F, the heavier hydrocarbons are cracked down to dry gas (mainly methane) which is normally referred to as (thermogenic gas). At greater depths where temperature exceeds the 450^0 F-level, all hydrocarbon molecules are destroyed and the host sedimentary rocks starts to be converted into metamorphic rocks.

3.2.6 The Thermal Maturation Curve

The phrase (thermal maturation) of kerogen is used to describe the change of Kerogen into hydrocarbon matter (oil and gas) by heating processes. Type and quantity of the produced hydrocarbon depends on the effective temperature which is dependent on the depth of the host sedimentary rocks. The Kerogen maturation process, with oil and gas generation, is function of depth of the source rocks. This is explained as follows (Stoneley, 1995, p. 33):

Kerogen starts to change into oil when temperature reaches about 150^0 F. The generated material at this stage is heavy oil. At higher temperature (attained at greater depths) the large hydrocarbon molecules break down (crack into) the smaller molecules light oil.

Common observations indicated that the generated liquid hydrocarbon (oil) is maximum at depth range around (3-5) km. At greater depths, where temperature increases to higher levels, less oil and more gas (wet gas) are produced. As depth increases to levels where temperature reaches about 350 °F, only dry gas (thermogenic gas), mostly methane, is generated. With further increase of depth, and at temperature exceeding about 450 °F, hydrocarbon molecules are completely destroyed, and the host rocks change into hard crystalline rocks (metamorphosed rocks), in which no any hydrocarbon matter is generated.

The process is normally expressed in the form of a curve (known as the thermal maturation curve), showing the changes of oil/gas proportions with increasing depth (Fig. 3.2).

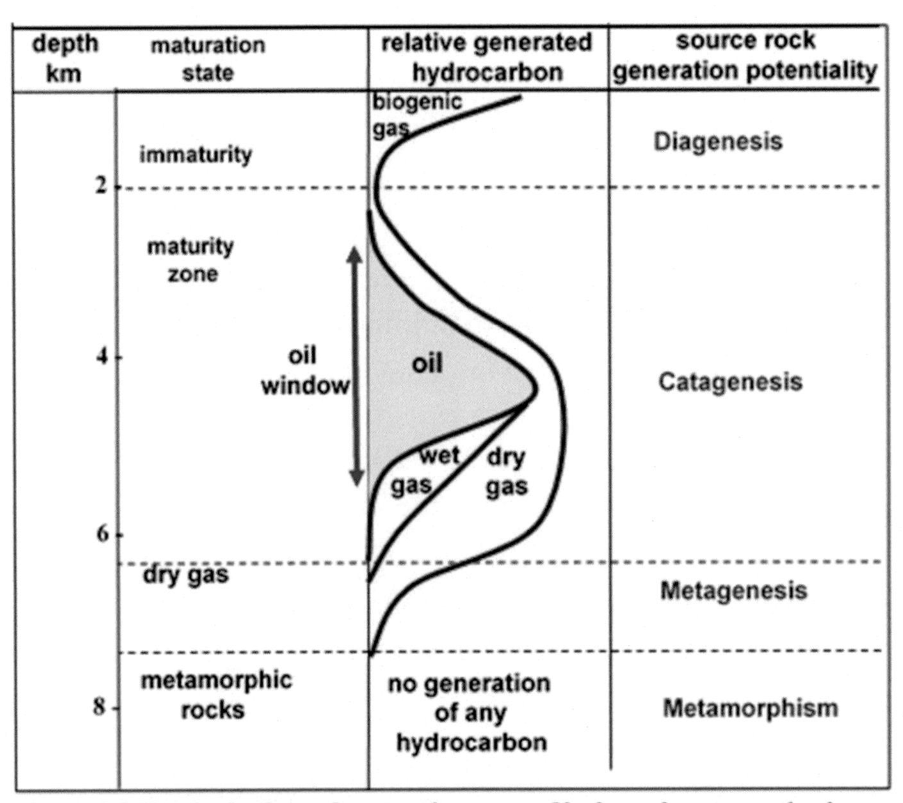

Figure 3.2 Sketch of a thermal maturation curve of hydrocarbon generation in the source rocks. Based on (Hobson, 1984, p. 15).

The general shape of the curve is the same for all source-rocks locations, but its detailed statistics vary from location to location. As it is apparent from the curve (shown above), oil is generated within an oil window of roughly (3-5) km. At depths greater than that of the curve-peak, the oil-to-gas ratio decreases. At depths greater than about (7) km, rock-metamorphism conditions prevail, and no hydrocarbon matter is expected to be generated.

There are several methods for determining the degree of maturation. One way depends on measurements of extent of reflection of light from the matter (Vitrinite) which exists with the hydrocarbons and coals. It was found that there is a direct proportionality between the (Vitrinite reflectance) and degree of thermal maturity. In general, the generated oil is darker when it is more mature. The colour is light yellow for low maturity, increasing in darkness with the increase of maturity.

3.2.7 Role of Pressure and Temperature

It is a common observation that both pressure and temperature are increasing as depth increases. In case of the presence of the organic matter, the Kerogen, the quantity and quality of the generated hydrocarbons depend on the pressure and temperature dominating the oil kitchen. The important feature of pressure and temperature effects on the generated oil and gas is that the proportion of the generated gas increases as temperature increases, and this ratio (gas/oil ratio) decreases as pressure increases. In other words, pressure has an effect on the gas/oil ratio which is opposite to that incurred by temperature.

Geological studies indicated that temperature is, in general, increasing with depth. Temperature depth-wise gradients in sedimentary rocks vary widely from place to place. Values as low as 5 °C/km, and as high as 90 °C/km, have been reported (Hobson, 1986, p. 13). In the geological literature, it is normally quoted that the geothermal gradient, measured down a bore-hole, falls within the range (25-30) °C per kilometer. With this rate of increase, it is expected that the temperature attained at the depth-range (3-5) km will be approximately equal to (120-180) °C or (250-360) °F, assuming surface temperature to be 30 °C.

3.2.8 The Mother Rocks

The rock medium, in which hydrocarbon is generated, is called (source rocks), and also, as sometimes appropriately called, (mother rocks). These rocks, within which oil is formed, consist basically of fine-grain sedimentary rocks (as shale, clay, and marl) which have been deposited under anoxic marine-deposition environments. Two main factors are governing the oil quantity produced by the mother rocks. These are: the organic matter, which were originally deposited with accumulated sediments, and the degree of maturation of the formed Kerogen. Both of these factors are, in turn, governed by the total geological environments.

3.3 Oil Migration

Geological studies ascertained that the oil generated within the mother rocks does not, as a rule, stay stagnant in the place where it is generated, but moves to the neighboring rock media which are of porosity and permeability that allow this process to take place. The hydrocarbon matter, in its liquid or gaseous state, leaves the mother rocks following a horizontal, vertical or along any other permeable pathway. This movement of the generated oil from its place of maturation is commonly referred to as (oil migration).

Naturally, the oil body needs some type of energy for activating the migration process. It is believed that the most effective factors (energy sources) are: pressure, buoyancy, capillarity, and the Earth gravity.

3.3.1 Factors Affecting Oil Migration

The oil migration phenomenon is the result of interaction of some types of potential energy which get converted into dynamic energy that forces the oil body to move outward from the generation location. The migrating oil keeps moving through the possible routes, then it, either gets blocked and prevented from further movement, or it reaches the surface forming seepage if the pathway allows.

Three groups of factors are governing quantity of the migrating oil and the pathways that are taken during motion. These are: migration environments, rock and oil matter properties (Fig. 3.3).

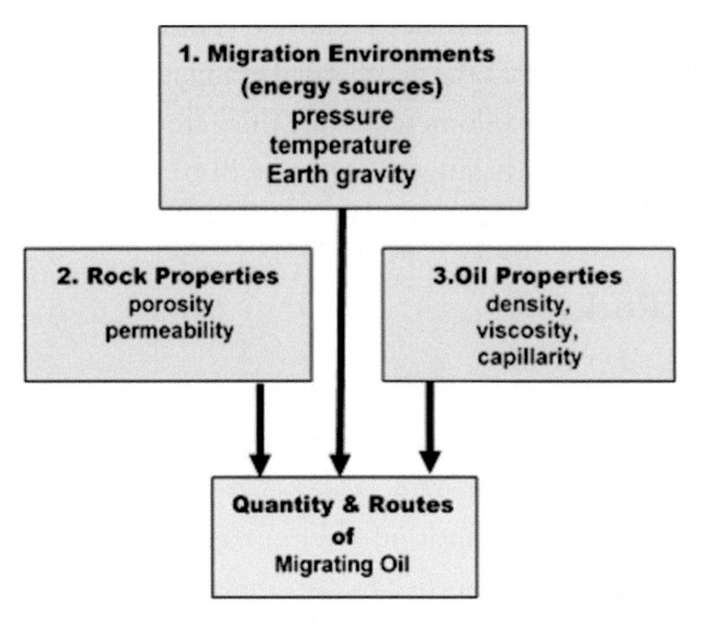

Figure 3.3 Main factors governing quantity and migration pathways of migrating oil.

3.3.2 Migration Environments

The main elements of the environments under which oil migrate from its generation zone, are pressure and temperature changes, in addition to the Earth gravitational force. In essence, these factors are the energy sources which cause the generated oil to move, that is, to migrate from the oil kitchen.

(i) Pressure Factor

The generated oil is under the effect of the pressure imposed on it from the overburden rocks. The severity of the effective pressure depends largely upon the total thickness of those overlying rock-layers. Two main types of pressure are known to be contributing to the resulting pressing force that causes the oil to migrate. These types are: hydrostatic and geostatic pressures.

The hydrostatic pressure is resulting from the quantity of fluids (oil-water mixture) existing at levels higher than the main oil body. A more effective pressure comes from the weight of the rocks accumulated over the generated oil. This type of pressure (known as geostatic pressure) is estimated to be more than double that of the hydrostatic pressure because of the higher rock density. Pressure is described as dynamic (e.g. geodynamic pressure), instead of static, when it is caused by an external force, as for example, when the host rocks, and the fluids within them, are involved in tectonic activities. Gas saturation pressure is another type of pressure that can contribute to the energy that causes oil to migrate.

(ii) Temperature Factor

The principal source of heat in the Earth crustal rocks is the Earth Mantle which is experiencing the phenomenon of convection currents created by the Mantle's heat changes. By the conduction principles, heat is transmitted from the Upper Mantle to the crustal rocks. In addition to that, heat is generated as a result of friction that occurs within rocks when affected by folding and faulting processes.

Generally speaking, Temperature changes cause some corresponding changes in the properties of fluids. Thus, for example, as temperature increases, oil volume, its viscosity, and pressure, increase, while oil surface tension and gas solution in the oil decrease. These effects incurred by temperature changes have corresponding effects on the oil movement.

(iii) Earth Gravity Factor

Due to the Earth gravitational field, any body of a defined mass is under a vertically down-pulling force which is of magnitude proportional to density of that body. Thus an oil body existing within a dipping rock-layer is under the effect of a gravitational force that has two components; one is perpendicular, and the other is parallel, to the rock layer. The "parallel" component tends to pull down the oil body in the dip direction by a force proportional to the dip magnitude.

Another type of effect of the gravity field is separation of co-existing fluids into water at the base, with liquid oil above it, and the gaseous hydrocarbon at the top.

3.3.3 Rock Physical Properties

The physical properties which have impact on oil migration are porosity and permeability. Porosity expresses the rock capability of containing fluids, while permeability provides the means for the fluid to move from place to place through the rock medium.

(i) Porosity

Porosity of a rock is defined as the ratio of the total volume of the voids in a given rock-body to the total volume of that body. This ratio is normally expressed as a percentage. It is called (primary porosity) when the porosity has been generated at the time the sedimentary rock has been formed, while it is described as secondary when it is created after the rock has been deposited. Sandstone rocks, for example, usually have primary porosity, whereas limestone rocks have, in general, secondary porosity, when they get fractured after being formed.

There are several factors affecting porosity. Particle sphericity, grain size, compaction pressure, and particle packing are factors which are affecting porosity. Two modes of packing are shown in (Fig. 3.4) where ideal grain shapes (spherical shapes) are used to calculate the resulting porosity in each mode of packing. It is found that in "cubic packing" and in "rhombic packing" the porosity values are about 48% and 26% respectively (Holmes, 1965, p. 414).

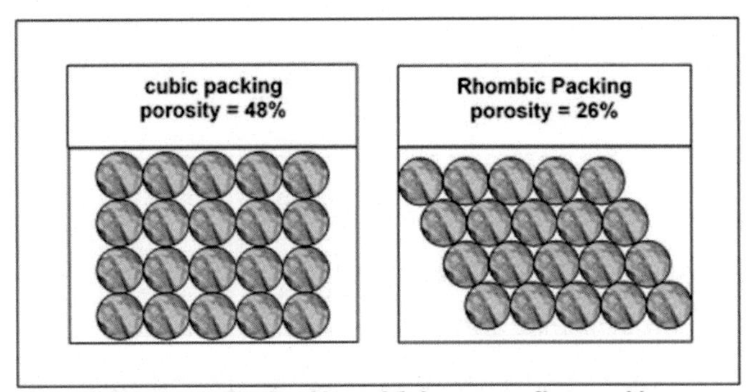

Figure 3.4 The two modes of packing and their corresponding porosities. Computations are based on the assumption that the packed grains are perfect spheres of equal sizes.

(ii) Permeability

Permeability is the second important physical property of rocks, after porosity. Permeability expresses the ability of a porous rock in allowing fluids (liquids and gases) to pass through it. In petroleum geology, it is mentioned that there is a general trend of increase in permeability with porosity. The relation is found to vary with type of rock lithology, grain size, grain sorting, diagenetic, and compaction processes. The important feature of a porous and permeable rock is that the rock contains void pockets (pores) allowing storage of fluids and those pores are interconnected allowing passage of the fluids through the rocks (Fig. 3.5).

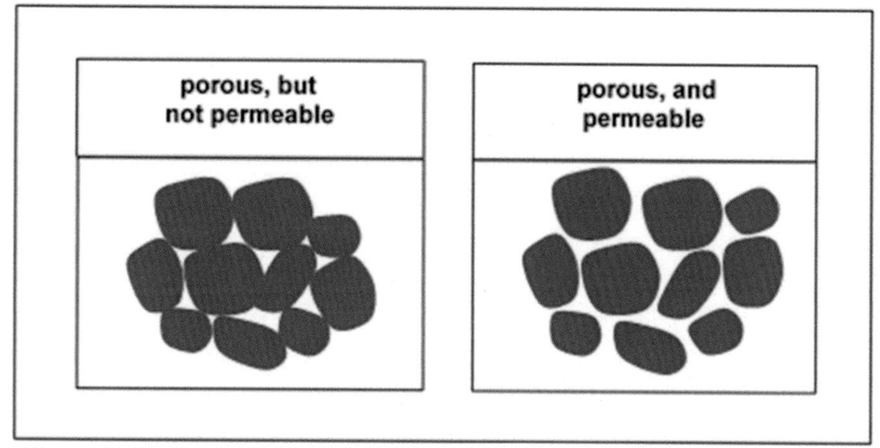

Figure 3.5 Porous rocks having grains with unconnected pores, whereas porous and permeable rocks are having connected pores.

Permeability is normally measured using Darcy's law, which relates the fluid flow-rate (**Q**) to its viscosity (**μ**), and the applied pressure gradient (**Δp**). If measurement is performed on a unit-cube of rock sample, and (**Δp**) is measured across two of its opposite faces, the coefficient of permeability (**K**) will be given by:

$$K = \mu Q/\Delta p$$

The measurement unit of permeability is called the darcy, after Henry Darcy (1803-1858). In practice, the smaller unit; the millidarcy, is more commonly used.

Most reservoir rocks have permeability coefficients in the range of (100-500) millidarcy (Selley, 1983, p. 34).

3.3.4 Oil Physical Properties

Naturally, the physical properties of oil have effective impacts on the oil motion through the porous and permeable rocks. Density, viscosity, and capillarity are the main properties which are controlling oil migration. .

(i) Density

Due to gravity force, an oil body is acted upon by a downward vertical force of magnitude that depends on the oil density. Another force associated with oil density is the buoyancy force which is proportional to the difference between oil density and water density. Thus, a free oil body found within higher-density water will tend to move vertically upwards by the buoyancy effect. Crude oils have densities falling in the range 0.75-0.9 gm/cc (Chapman, 1976, p. 28).

(ii) Viscosity

The fluid viscosity is an expression for the fluid resistance to flow when it is under some pressure difference. It is usually measured by the unit (dyne.second/cm^2) which is called (Poise) after the French physicist Jean Poiseuille (1799-1869).

(iii) Capillarity

In general, surface tension of liquids causes the liquids to move across rock pores in the same way as it moves through a capillary tube immersed in that liquid. This phenomenon is a well known physical property, called (capillarity).

Capillarity of a liquid depends on the nature of liquid (oil or water) and on the size of the rock pores. The resulting capillarity pressure is inversely proportional with pore size, and the motion is expected to be faster in rocks of smaller pores.

There are still other factors affecting liquid motion in rocks. Diffusion property (expressing liquid motion depending on degrees of concentration) is one of such factors. Oil moves in the direction where it is of lower concentration. Another property is the liquid (wetability) which describes the extent of spread of a liquid on the surface of a solid in contact with it. These properties have minor effects compared with the above mentioned three properties (density, viscosity, and capillarity).

3.3.5 Mechanism of Oil Migration

Petroleum studies showed that the generated oil is, in general, expelled from the source rocks out to the neighboring porous and permeable rocks which allow oil to pass through and continue its movement horizontally, vertically or through any other possible permeable path. The energy required for the migration is provided by interactions of several factors. Most important of these are overburden pressure, capillarity pressure, gravitational force, and buoyancy of the oil in respect to the associated formation water.

Mechanism of oil migration is based on the assumption that oil droplets under pressure move through microscopic fracture supposed to be existing in the source rocks in the direction of the local pressure gradients. Movement of oil is believed to occur in two stages. At first, oil is expelled out of the generation habitat (primary migration), then movement is resumed through pervious rocks (secondary migration) until reaching a closed porous zone (oil trap) in which oil is accumulated. In cases where entrapment conditions are not met-with, oil eventually finds its way to the earth surface forming an outlet, commonly referred to as (oil seepage).

3.3.6 Stages of Oil Migration

After generation, and under the thermal maturation processes, the hydrocarbon matter is settled down in the generation zone (the oil kitchen) in the form of microscopic droplets collected with the interstitial water that was captured from the sea water in which the precipitation process has occurred.

As it is mentioned above, the so-formed oil starts to move away from the source rocks (mother rocks) by certain energy factors. Although details of this process (oil migration) is not yet fully understood, but petroleum geologists have agreed that oil motion from the mother rocks to the accumulation zone occurred in two stages; the primary stage (oil transfer from mother rocks to surrounding rocks) and the secondary stage (oil transfer from the surrounding rocks to entrapment structures).

(i) The Primary Migration
Transfer of oil from the impervious mother rocks to the neighboring pervious rocks is normally referred to as Primary migration. This process requires that there is some sort of pathways in the mother rocks which allow passage of the moving oil. To explain the process, it is assumed that the impervious mother rocks are having microscopic fractures which have allowed oil, and over millions of years, to migrate to the surrounding porous and permeable rocks.

(ii) The Secondary Migration
In this stage oil flows through the porous and permeable beds, reached at by the primary migration, and continues to move until it meets some barriers which stop the reaching oil from any further motion. In these entrapment structures oil accumulation takes place forming what is called an oil reservoir. The other alternative is that oil continues in its motion until it reaches the surface where it forms an oil-seepage.

3.3.7 The Pitch Lakes and Rock Oils

During oil migration from the mother rocks to the seepage vent, the hydrocarbon gases find their way to the atmosphere through that vent. Under these conditions, the left-over heavy oil (or thick viscous tar matter) spread out on the surface in the neighborhood of the seepage zone in the form of solidified sheets of asphaltic deposits. Examples of such deposits are: the Pitch lake of Trinidad and the surface tar deposits found in the Marij area near Hit town in Alanbar Governorate in Iraq. The Marij tar deposits seem to have been used by ancient

Iraqis (Sumerians) as mortar and road paving materials about 5500 years ago. (Holmes, 1965, p. 458).

Under special conditions, some shallow rock layers become impregnated with bituminous deposits of mainly type-3 Kerogen. On heating, these rocks are producing hydrocarbon liquids, normally referred to as (rock oils). Examples of the rocks which can be used as sources of oil are: the Athabasca oil sands found in northeastern Alberta in Canada and the Colorado (USA) oil-shale deposit which is believed to be the most concentrated hydrocarbon deposit in the World. For mainly economic reasons, production of oil from these hydrocarbon-bearing rocks has lagged behind the conventional deep-source oils.

Chapter-4

4. OIL ACCUMULATION

Due to differential pressure and other natural energy sources, oil is forced to move (migrate) through permeable rocks until reaching a barrier against which oil stops and begins to accumulate in the rock pores forming what is called (the oil reservoir). The accumulating fluids (mixture of water, oil, and gas), get sorted out into separate bodies of water, oil, and gas. The sorting process occurs mainly because of the differences in their respective densities. Because of this phenomenon (known as buoyancy), the mixture is sorted typically into a separate oil body with water below it, and gas above it.

4.1 The Petroleum System

The petroleum system is a comprehensive term that covers the characteristics of the hydrocarbon accumulation and all associated processes involved in the generation, maturation, migration, and its final destiny, accumulation in a closed reservoir rock (oil trap) or dissipation on the Earth free surface (oil seepage).

The principal elements of a petroleum system are: source rock, reservoir rock, sealing rock (or cap rock). These elements make up the essential conditions for securing entrapment of the hydrocarbon matter. The principal aim of evaluation of a petroleum system is determination of its potentiality in having oil and gas in commercial quantities. These elements are briefly defined as follows:

(i) Source Rock

As their name indicates, source rocks represent the rock medium in which hydrocarbons were formed including the whole range of processes starting with the organic deposits, their maturation into hydrocarbon matter, and migration (primary migration) to the neighboring pervious rocks. Properties of source rocks (normally made of fine grain argillaceous rocks as shale, clay, and marl or limestone) are investigated mainly by geochemical analyses.

(ii) Oil Maturation

The first step in oil generation is transformation of the preliminary organic matter into the more homogeneous matter (the Kerogen) which, in turn, gets converted (wholly or partially) into hydrocarbon matter (oil and/or gas). As it is mentioned above, the main factors controlling this process (maturation process) are mainly pressure are temperature, both of which are dependant on the depth of the oil kitchen.

(iii) Oil Migration

After its generation, the hydrocarbon matter migrates from the source rock to the reservoir rock in two stages. In the first stage (primary migration) oil is expelled from the source rock into the near-by pervious rocks, and then moves (secondary migration) to the oil trap, which is the reservoir rock-formation. The factors controlling the migration process are rock compression, the buoyancy phenomenon, gravitational force, and volume changes due to chemical and thermal changes.

(iv) Reservoir Rock

These rocks are essentially porous and permeable rock formations through which oil and gas are migrating (secondary migration) until reaching an entrapment geological structure where accumulation takes place, forming a hydrocarbon reservoir. Typical reservoir rocks are sandstones and fractured-limestone formations.

(v) Oil Trap

An oil trap is a confined space of porous and permeable rocks within which oil is accumulated. An oil trap is a geological feature that has the following characteristics:

- Reservoir rocks which are considered to be the core of the oil trap. It is the oil-bearing porous and permeable rock formation

- Seal rocks (or cap rock) which is made of impervious rocks which are capping the oil-bearing reservoir rocks, preventing oil from escaping from the reservoir rocks of the oil trap. Sealing efficiency depends on the extent of lithological tightness that controls the hydrocarbon retention in the reservoir rock formation. Tight limestone (as chalky limestone), shale, or evaporites are typical lithologies of a sealing formation.

- Geometrical shape and dimensions (extension and thickness) are necessary characteristics of an oil trap capable of holding oil accumulation in quantities adequate of commercial exploitation.

There are three main types of oil traps; structural (as folds and faults), stratigraphic (as sand lenses and coral reefs), and structural-stratigraphic combination traps (as salt domes and unconformities).

4.2 The Oil Reservoir

An oil reservoir is a subsurface pool of hydrocarbon fluids contained in porous rock body. Two main conditions must be fulfilled, in order that an oil reservoir is formed. Firstly, the rocks must be of sufficiently good porousity to allow containment of the incoming oil, and secondly, the rocks must be permeable to allow oil-movement within it. For this reason reservoir rocks are mostly made up of sedimentary rocks (sandstone or limestone). Within the bulk of such types of rocks, oil exists in the form of microscopic droplets mixed with the interstitial water filling pores of the rocks (Fig. 4.1).

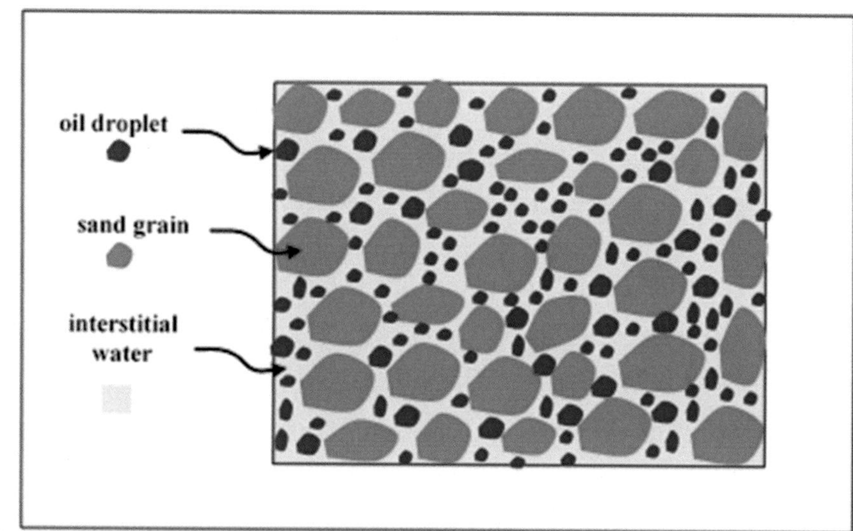

Figure 4.1 Schematic presentation of an oil-bearing sandstone reservoir rock.

4.2.1 The Oil Accumulation process

Through migration routes the hydrocarbon matter may move in the form of gas, oil vapors, oil condensates, liquid within which gas is dissolved, or heavy, high-viscosity oil. According to the prevailing pressure and permeability conditions, oil droplets (or gas bubbles) keep moving until meeting an entrapment space (oil trap) in which movement is stopped and the arriving fluids (gas, oil, and water) oil starts accumulating in that space, forming the oil reservoir. Due to the buoyancy phenomenon, the three constituents get separated from each other forming three distinct zones of water at the bottom, oil, in the middle, and gas at the top (Fig. 4.2).

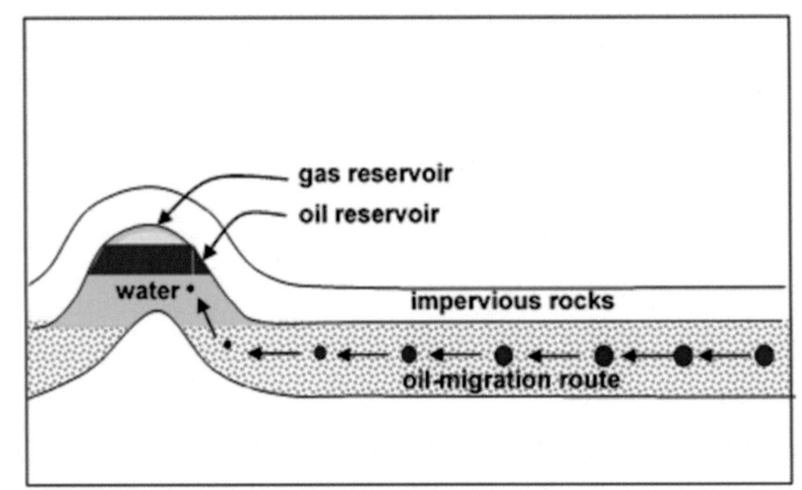

Figure 4.2 Oil accumulation process and formation of the oil reservoir.

In order to be economically rewarding oil-reservoir, the oil column must be of several meters and more, and the rock effective porosity to be no less than 10% with an adequate permeability that allows the hydrocarbon fluids to move freely to, and within, the entrapment space.

4.2.2 Fluid Sorting in a Typical Oil Reservoir

A typical geological structure of an oil reservoir is a geological dome or an anticline. The accumulated fluids in the reservoir rocks are normally sorted out into three distinct zones depending on their individual densities. The hydrocarbon gas forms a gas cap at the top of the reservoir, below which, liquid oil occupies a central position between the gas cap at the top and water (called formation water) below.

The gas-oil and oil-water contact surfaces (normally horizontal planes), are called gas-oil contact (GOC), and oil-water contact (OWC) respectively. The difference in elevation between GOC and OWC represents the oil column and the point at which oil starts to spill out of the trap-structure is called (spill point) which is a structurally lowest point below which oil shall leak out (spill) from the reservoir rock body. The contour line that passes through the spill point is called the (oil closure contour). A typical oil-reservoir of an anticlinal structure with its reservoir statistics are shown in (Fig. 4.3).

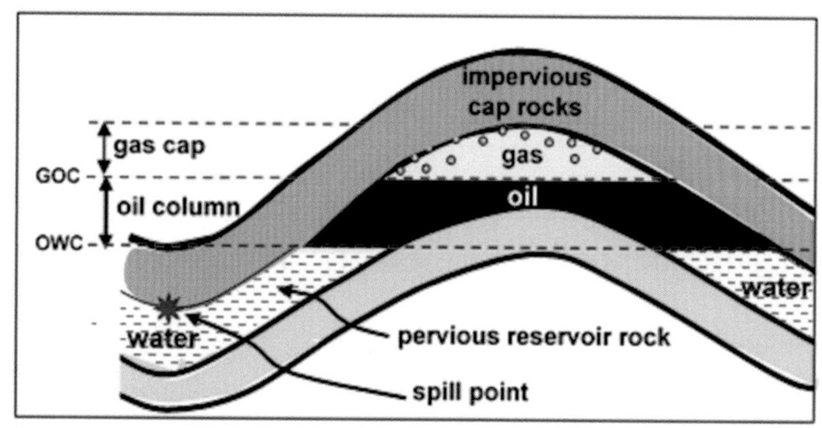

Figure 4.3 A typical hydrocarbon reservoir (geological dome).

The gas cap consists of low molecular-weight hydrocarbon of mainly paraffin-series type. The principal constituent is Methane, with smaller percentages of Ethane, Propane, Butane, and other paraffinic, and inorganic gases as carbon dioxide, Nitrogen, and hydrogen sulfide. Below the gas cap is the oil body which is saturated with the dissolved gas. The formation water, existing below the oil, is normally saline water containing certain percentages of salts as chlorides and sulfates of Sodium, Potassium, Calcium, and Magnesium.

Regarding the gas-oil contact (GOC) and oil-water contact (OWC) surfaces, it is found that they are usually horizontal planes, though they are, in certain cases, slightly inclined. In practice, the changes from gas to oil and from oil to water are not sudden changes but gradual changes. In fact, the oil body is separated from the formation water below and from the gas cap above, by transitional zones and not by clear-cut boundaries (Fig. 4.4).

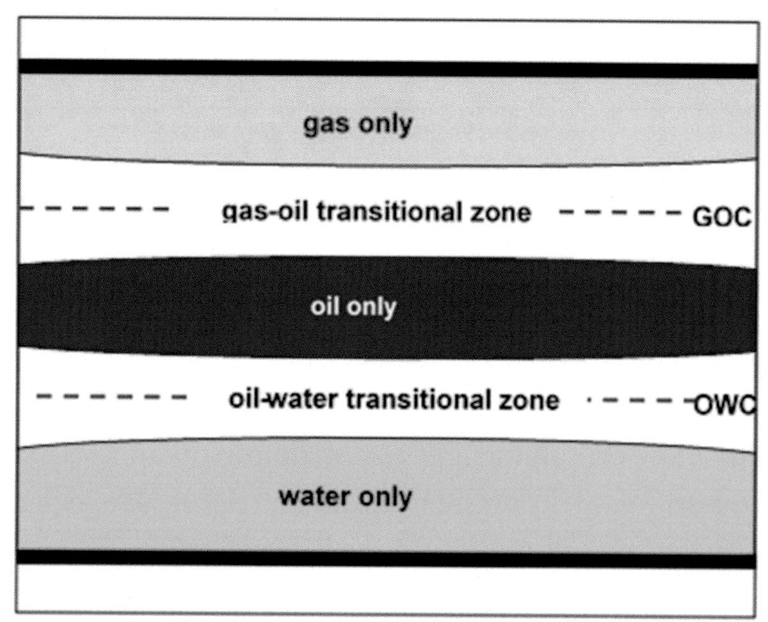

Figure 4.4 The oil body, in a reservoir, is separated from the gas cap and from the formation water by transitional zones.

A typical oil reservoir contains the three types of fluids (gas, oil, and water), however there are reservoirs with no gas cap (oil-only reservoirs) and others with gas only (gas-only reservoirs). In such cases only one contact surface exists, OWC for the case of oil reservoirs, and GWC for the case of gas reservoirs.

4.2.3 Depth and Thickness of Oil Reservoir

The first oil well, drilled by Edwin Drake in Pennsylvania, USA, in 1859 has produced oil from a depth of 21.2 m. In fact oil may be found as seepage on the surface or at depths reaching few kilometers below surface. Most common depths are 2 to 3 kilometers, and oil condensates or free gas are usually found at greater depths.

Thicknesses of the reservoir formations range from few meters to few hundreds of meters. Oil can be found in one thick sedimentary formation or distributed among a number of separate thin beds, as oil-bearing sandstone beds separated by shale beds. In such a case, oil in the multi-formation reservoir can be produced by one drilled well that penetrates all the oil bearing formations (Fig. 4.5).

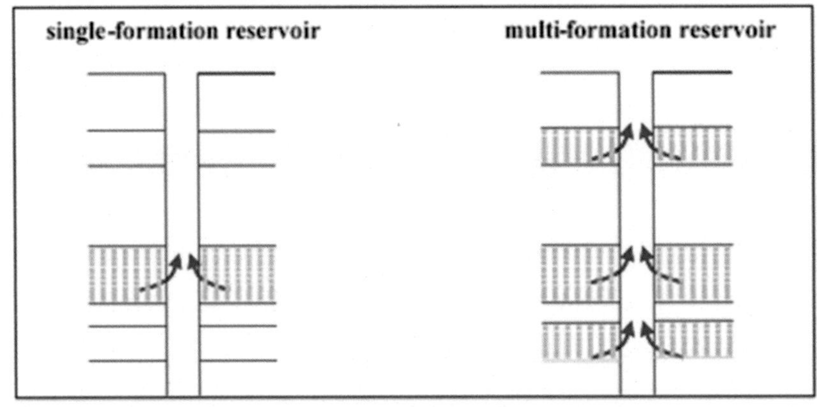

Figure 4.5 Single-formation and multi-formation reservoirs.

In general, oil reservoirs are found in formations which belong to Mesozoic and Tertiary Eras.

4.3 The Reservoir Rocks

Two main conditions must be fulfilled, in order that an oil reservoir is formed. The reservoir rocks must be porous to allow containment of the incoming oil and permeable to allow free oil-movement. In short, reservoir rocks must be porous and permeable rock media. With these properties, almost all known reservoir rocks are made up of sandstone or limestone types of sedimentary rocks.

4.3.1 Sandstone Reservoirs

Sandstone reservoir rocks are generally characterized by various degrees of porosity and permeability. Such rocks are rated as good oil reservoir rocks when they are made up of clean sands (with no fine grain sediments, as silts or clay). Porosity improves as the sand grains are large and nearly-spherical in shape. Further, the sand grains should not be too tightly cemented as it happens in cases where the sedimentation has taken place in water saturated with silicate or carbonate matter.

In addition to the role of the sedimentary environments, other effective factors may come into play after the deposition process. With increase of depth of a given formation, compaction pressure increases, with the result of reducing the porosity and permeability of that formation. Under certain chemical conditions, some of the rock particles are dissolved by certain chemical processes (digenesis processes) bringing about increase of the voids between the rock particles and hence improvement of porosity.

Since sedimentary depositional environments are normally changing from place to place, it is expected that reservoir size, type of lithology (rock facies), and other reservoir characteristics are accordingly changing. As far as oil accumulation is concerned, facies changes create favorable conditions for oil trapping. Thus, for example, change of clean sand deposits (silt-free or clay-free sandstone) into shale or clay can provide both of the reservoir formation (sandstone reservoir) and the seal formation (shale or clay seal-bed). It is left for the exploration activities to determine the effective volume of the oil reservoir and physical rock properties.

4.3.2 Carbonate Reservoirs

The second well-known type of reservoir-rocks, are the carbonate rocks which are deposited in marine chemical environments and consisting mainly of Calcium carbonate (forming limestone) and sometimes with Magnesium (forming dolomite). Other sources of these rocks are skeletons of corals and sea shells of some marine crustacean livings which are normally characterized by better porosity and permeability properties. As with sandstone reservoirs, carbonate reservoirs rocks undergo changes with time due to the effects of depth-dependent compaction pressure, fracturing, diagenesis, and chemical replacement process that produces dolomite rocks. Some of these processes decrease porosity as compaction pressure, and others increase porosity as fracturing and diagenesis.

The most important factor affecting reservoir oil potentiality are the primary porosity established when the rocks are formed, and the secondary porosity which is created as result of later processes, like rock fracturing, leaching, or chemical-replacement.

4.3.3 Effects of Rock Types on Oil Properties

Common observations indicated that oils, contained in reservoirs of calcareous rocks (as limestone) and reservoir-rocks containing sulfur in their chemical composition (as gypsum and anhydrite), would be of asphaltic-base oils with certain percentage of sulfur. Further, it is noted that oils produced from reservoirs of sand-shale sequences are paraffin-base, sulfur-free oils.

4.3.4 Reservoir Internal Conditions

Two main factors are governing the reservoir fluids; these are the pressure and temperature. The oil body is under pressure from the gas cap from above and the formation water from below. Increase of reservoir pressure leads to increase of gas dissolution in the oil until a state is reached whereby the pressure (called the saturation pressure) can no longer be able to get dissolved and at this point a free gas body begins to accumulate to form the gas cap.

An equally effective factor is the temperature which has direct effect on both the gas and the oil below. With increase of temperature, volume of the hydrocarbon body (oil and gas) increases. This also leads to increase of the overall fluid pressure as well as increase of salt solution. At the same time, oil viscosity and gas solution in oil decreases with increase of temperature.

When an oil reservoir is provided with an outlet for the reservoir oil or gas to escape through (through a fault or through a drill-hole), hydrocarbon moves through the outlet by the reservoir energy which is created from the prevailing reservoir pressure. The reservoir energy is established from the associated gas (dissolved and free gases) and from the formation water pressure feeding the reservoir. Naturally, reservoir energy declines with oil depletion (as with oil production) leading the oil body to decrease with the consequence of falling down of the GOC and to rising of the OWC. There are some measures applied to restore the reservoir energy and enhance oil recovery as with gas- or with water-injections and other energy-boosting techniques.

4.3.5 Reservoir Energy

Reservoir energy is the potential energy responsible for moving the reservoir fluids in the direction of an outlet (as a drilled well) when that outlet allows free passage of the fluids. Sources of the energy are:

- Fluids pressure as that of the formation water found underneath the oil body.
- Free-gas pressure of the gas-cap found above the oil body.
- Saturation pressure of the gas found dissolved in the oil body.

- Compaction pressure due to the overburden rocks.
- Hydrostatic pressure due to the Earth gravitational field.

Briefly stated, the reservoir energy is resulting from within the three constituents of the oil reservoir; the free gas at its top, the liquid oil, and the formation water at the bottom. When oil production starts reservoir energy begins to decrease. As a consequence to continued production, pressure of the dissolved gas continues to decreases until reaching the level of saturation pressure when the dissolved gas starts to be liberated in the form of gas bubbles which eventually forms the gas cap above the oil body.

4.4 Oil Entrapment and Oil Traps

Migrating oil keeps moving so far as there are passages for it to move through. This is facilitated by the porous and permeable nature of the rock medium. Oil is stopped and gets collected when the moving oil meets a place having the required entrapment conditions. These conditions are represented by a reservoir rock that is confined by a sealing material.

A typical oil-trapping structure (oil-trap) is a geological dome in which the folded pervious reservoir beds are covered by an impervious bed that acts as seal preventing oil from escaping. In nature there are several types of subsurface oil traps which hold oil and gas. Common types of seals are shales, clays, evaporates (salt, gypsum, anhydrite). Impervious limestones (limestone rocks having no fractures) can also serve as seal, but not so common.

4.4.1 Types of Oil Traps

An oil trap is a subsurface geological structure capable of holding hydrocarbon accumulation in commercial quantities. In order to have a trapped oil-body, with no any dissipation, three conditions must be fulfilled. These are: generated oil from source rocks, pervious reservoir rocks for oil to accumulate in, and impervious cap rocks acting as a seal to prevent the oil from escaping.

Based on the type of geological activity involved in creating the trapping geological structure, oil traps may be classified under three different groups; structural, stratigraphic, and structural-stratigraphic combined traps.

4.4.2 Structural Oil Traps

The main geological structural activities affecting rock formations are folding and faulting. Folding can result in forming oil traps as anticlines and domes in such a way as to have the oil contained in the pores of the folded reservoir rock and impervious cap rock preventing the oil from escaping. Like folding, faulting can create a structural oil-trap. If it happened that a porous formation is shifted by a faulting process to a position whereby it becomes adjacent to an impervious layer which acts as a seal preventing hydrocarbons from moving through, a trapping structure is created.

(i) Fold Traps

Folds are created as results of natural forces acting on the crustal rocks forming various types of bed deformations including the favorable geometrical forms which suit oil trapping. Some of these folds are caused by compression forces due to tectonic or non tectonic forces. A folded bed may be due to an uplift (as that created by an igneous plug or by a salt diaper), or due to a drape phenomenon (as that caused by basement palaeo-topographic highs or over a basement horst block).

Typical examples of the fold traps are the anticlines of various types and sizes. Some of these are symmetrical and others are asymmetrical anticlines. An oil-trap formed in the form of a symmetrical anticline is shown in (Fig. 4.6).

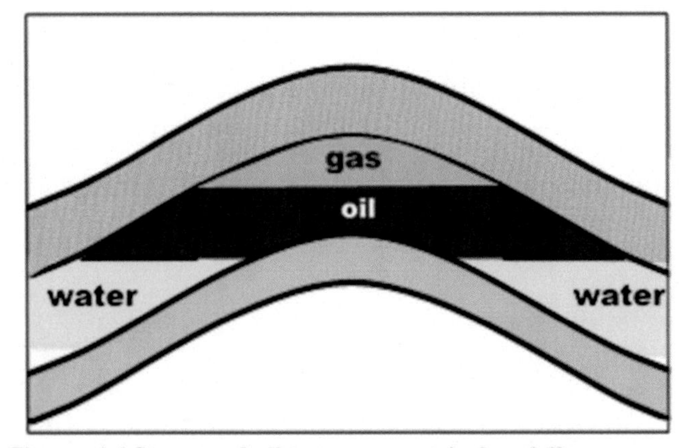

Figure. 4.6 Structural oil trap, symmetrical anticline

In the case of asymmetrical anticlines, the crest point of the involved beds, are shifted sideways by an amount of shift that increases with depth. This is in contrast to the symmetrical anticline where the crest points of the beds keep at the vertical axial line. Flank dips may change with depth, and in this case, trapping capacity may be changing from bed to the other within the anticline trap.

(ii) Fault Traps

Like folding, faulting can create a structural oil-trap. If it happened that a porous formation is shifted by a faulting process to a position whereby it becomes adjacent to an impervious layer, acting as a seal preventing hydrocarbons from escaping, a trapping structure is created.

Most common types of faults which contribute to creating oil traps are the normal faults due to tensional forces, reverse faults due to compression forces, tear faults (wrench faults), and growth faults which are contemporaneous with the precipitation process. A typical fault trap is shown in (Fig. 4.7).

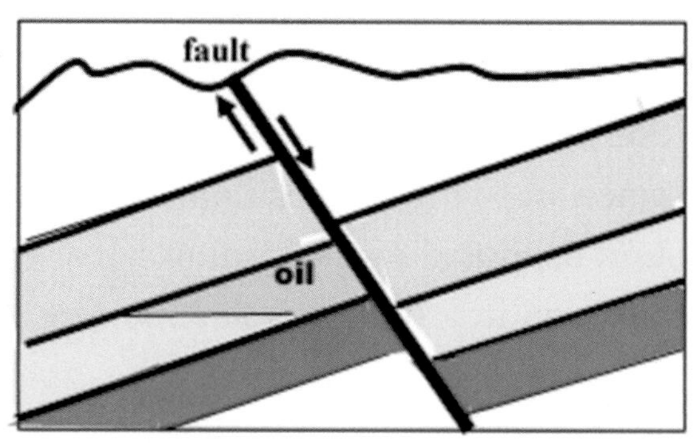

Figure. 4.7 Structural oil trap, normal fault

Presence of a fault in a set of geological formations may be creating a trap or destroying a trapping mechanism. A layer displacement, caused by faulting, may result in a state whereby an oil-reservoir (wholly or partially) becomes juxtaposed with a porous and permeable formation found on the other side of the fault plane. In this case, oil escapes through the porous pathway and no trapping conditions exist. Another possible negative aspect may be provided by the fault and that is by the escape of oil through the fault zone, which acts as a passage for oil to move through. When fault plane extends to surface, seepage may be formed.

4.4.3 Stratigraphic Oil Traps

A stratigraphic oil trap is formed as a result of a stratigraphic activity like sediment deposition, diagenesis (rock physical and chemical changes at relatively low temperature). As a result of these changes, vertical and lateral changes may take place in bed thickness, lithology (facies changes), and rock porosity. Sand-to-shale transformation is one of the principal features which can lead to a stratigraphic trap. Secondary stratigraphic traps, developed after lithification of deposits, are mainly due to diagenesis. This process may lead to porosity improvement (due to dissolution) or porosity deterioration (due to cementation).

Reservoir rocks in stratigraphic traps may be of sandy or carbonatic nature. Traps may be formed from deposition of sandy material in confined depressions or sand-filled channels formed in fluvial or deltaic channel environments. Corresponding to the sand-based reservoir rocks, there are the carbonatic deposits as those formed by remains of the coral reefs. In general, the stratigraphic traps, found by exploration activities, are far less than the structural traps, as they form about 13% of the world-wide discovered oil and gas fields (Halbouty, et al., 1970).

Typical examples of stratigraphic traps are sand lenses, buried channels, and coral reefs.

(i) Sand-Lens Traps

Sand lenses and (sand-filled channels) are stratigraphic features which have been formed in fluvial or deltaic deposition environments, as fluvial or deltaic channel sands. Similar sedimentary features are formed in marine deposition environments, as the marine (sand bars) which are normally found deposited in the form of bars extending along lines parallel to palaeo-shorelines. A cross-section in a typical sand lens, deposited within shale medium, is shown in (Fig. 4.8).

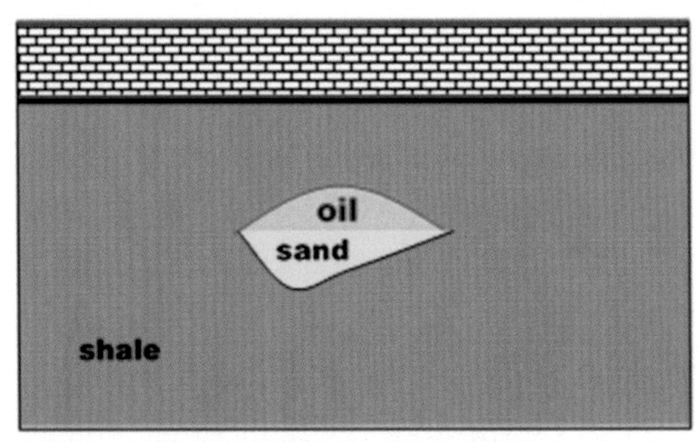

Figure. 4.8 Stratigraphic oil trap, sand lens

By nature, sandstone is usually characterized by good porosity and permeability, which makes sand lenses embedded in non-porous rocks (as shale or clay) makes the appropriate environment for oil entrapment. A sand pinch-out lying against a basement high can form an oil trap of this type.

(ii) Corral-Reefs Traps

. Reef deposits are characterized by good primary porosity and secondary porosity resulting from certain chemical changes as dissolution and replacement chemical processes (digenesis processes). Like sand lenses reef deposits, found deposited within shale or clay, can become good stratigraphic oil traps (Fig. 4.9).

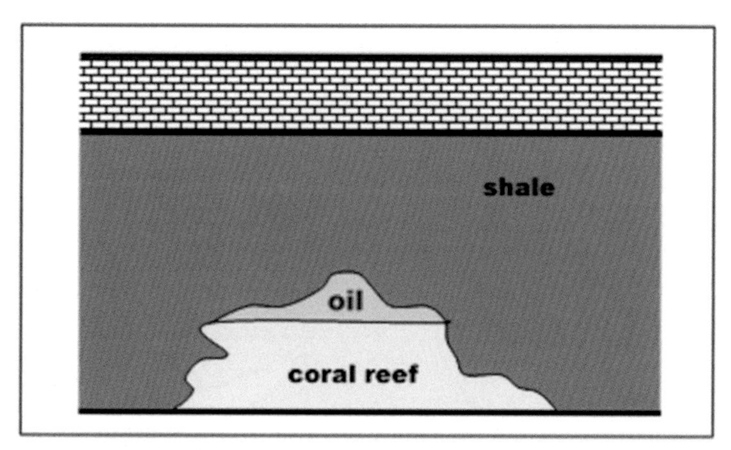

Figure. 4.9 Stratigraphic oil trap, coral reef

4.4.4 Structural-Stratigraphic Combination Traps

Some oil traps were formed as a result of combined effects of both structural and stratigraphic activities. Salt domes and angular unconformities are typical geological features in which both of structural and sratigraphic elements enter in their generation mechanism. On the basis of the geologic structural and stratigraphic processes involved in their creation, oil traps associated with salt domes and unconformities are classified as structural-sratigraphic combination traps (or just combination traps).

(i) Salt Domes Traps

Under certain geological conditions, subsurface salt bodies (known as salt plugs or salt domes) grow upwards pushing the overlying sedimentary strata causing them to get folded, faulted, and fractured. Very often the overlying layers are pierced by the growing salt body. The upward motion of salt bodies pushes the overlapping strata. This process, associated with deformations (layer tilting, folding, and piercing), is termed salt-tectonic or (halokinetic) process.

Growth of a salt body is a stratigraphic phenomenon, but the folding of the pinch-outs created by its piercing effect, represents the structural element. Being impervious material, the salt dome acts as an effective seal for preventing oil found in a reservoir layer from escaping.

A salt dome, with the associated halokinetic deformations, is shown in (Fig. 4.10).

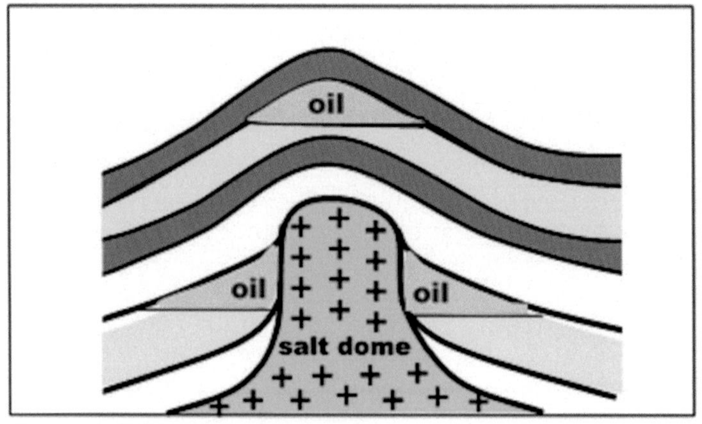

Figure 4.10 Structural-Stratigraphic combination trap, reservoir porous bed pierced by a salt dome.

- Folded of the overlying sedimentary layers (structural traps).
- Pinch-outs of strata formed against the salt dome (stratigraphic traps).
- Pierced strata which curve upward against the salt body (combination traps).

(ii) Unconformity Traps

An unconformity oil-trap is formed when a porous formations is covered by a younger non-porous formation having a relatively different dip angle. It is considered to be of a combined-type, because of the folding and tilting of the older set of formations (structural elements) followed by erosion and re-deposition (stratigraphic elements) that follows the folding process. An unconformity oil-trap is formed when the tilted oil-bearing formation is truncated by the unconformity surface separating it from the unconformable impervious rocks as shale or clay formation (Fig. 4.11).

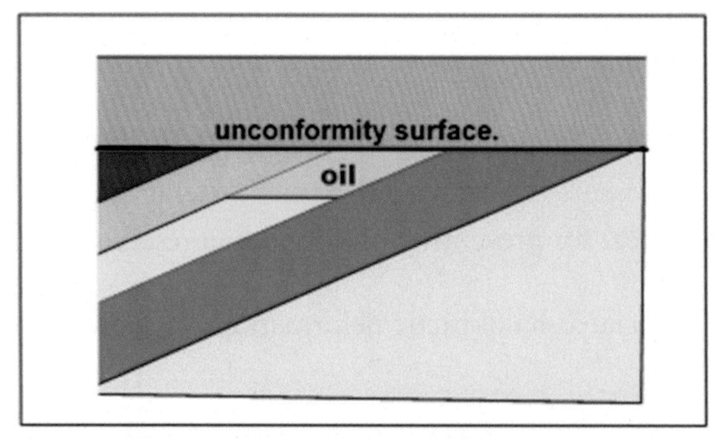

Figure 4.11 Structural-Stratigraphic combination trap, reservoir porous bed cut by an unconformity surface.

4.4.5 Hydrodynamic Traps

There are some other types of traps which are far less common than the types cited above. One type of traps (called hydrodynamic traps) is caused by the differential pressure of the water associated with the oil reservoir (Selley, 1983, p. 59). When pressure is strong enough, the oil-water contact becomes tilted rather than being horizontal as it is in normal cases (Fig. 4.12).

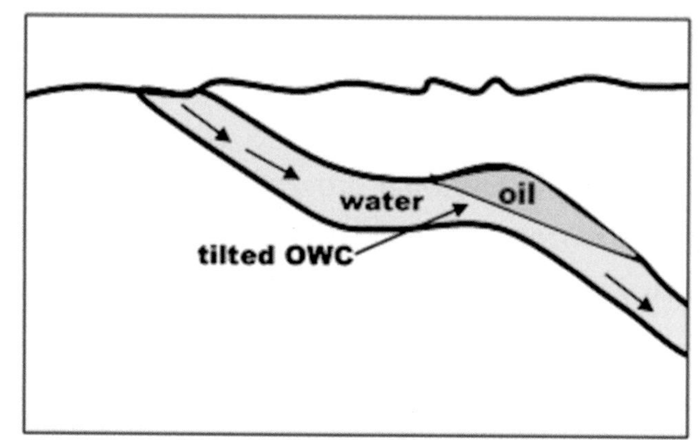

Figure 4.12 Structural-Stratigraphic combination trap, hydrodynamic trap causing tilting of the OWC.

4.4.6 Oil Seepages

In cases where the migrating oil finds its way to the earth surface, an oil-seepage is formed. This phenomenon takes place when reservoir rocks are outcropping or when there is an open pathway connecting the reservoir rocks to the surface. An oil seepage also occurs with the existence of fracturing or faulting in the sealing impervious formations of the oil trap. Oil seepages serve as direct indication for presence of possibly large oil reservoir in a subsurface trapping structure. Of course, the oil reservoir feeding the oil seepage is not necessarily located vertically below the surface location of the seepage. Geological knowledge and other geophysical exploration-means can help to determine the actual location of the feeding oil reservoir.

An oil-seepage may take the form of gas or oil spring or even hard hydrocarbon matter. The asphaltic layers, found in nature as surface deposits, are being formed from oil seepage after losing the volatile constituents by normal evaporation process that has taken place over past geological times. The surface asphalt deposits found in the city of Hit (Alanbar-Iraq) is an example of an oil seepage that formed an asphalt surface layer after loosing the volatile constituents. The oil well (Hit-1), drilled in 1938 near-by the asphalt area, proved the existence oil. The test production of this well showed a production rate of 5000 b/d of heavy oil with relatively high sulfur content.

4.5 Role of Earth Orogenies in World Oil Accumulations

Continental blocks, throughout geological history, acted as sources of sediments feeding major depressions to form large sedimentary basins. Due to the on-going tectonic movements and other geological processes, these basins suffered from various structural and stratigraphic changes. Oil accumulations, which are mostly associated with sedimentary basins, are naturally affected by the Earth major orogenic movements.

4.5.1 The Earth Major Seismic Belts

Geologists have identified three major global orogenies. These are: the Caledonian orogeny which occurred during the Palaeozoic era, the Hercynian (Variscan) orogeny by late Palaeozoic era (Carboniferous-Permian), and the Alpine orogeny which is believed to have started in early Tertiary, when its major phases of mountain building occurred in the Paleocene-Eocene periods, and continued up to present times.

According to geological knowledge, the Caledonian and Hercynian movements have created the Scottish mountains, the Appalachian mountain ranges (North America), and the Ural Mountains in the Eastern Europe. The depression of the Mediterranean-Sea region (known as the Tethys depression), is believed to be of Hercynian origin. The oil fields found to the west of the Urals and those found in areas surrounding the Appalachians are considered to be of Palaeozoic age and their oil-traps are related to Caledonian and Hercynian orogenic movements.

The latest and most important orogenic movement is the Alpine movement. It has created the Alpine mountain ranges (the Alpide Belt) which extend from Atlas Mountains in North-western part of Africa, passing through Europe (Pyrenees, Alps), the Middle East, to the Himalayas and continue eastward through the South-Eastern parts of Asia to Indonesia and Philippine where it joins the other seismically-active belt, the Circum-Pacific Belt. The Rocky Mountains and the Indies in the American continents are included in these Alpine mountain ranges which are generally known by their seismic and volcanic activities, (normally referred to as the seismic belts).

4.5.2 The Earth Major Oil Belts

In general, oil traps are found in association with the sedimentary basins which were involved in these global movements. The greatest part of the world oil-traps are found in geological structures associated with the Alpine movement. The oil fields in North Africa (associated with the Atlas Mountains), the European oil fields (associated with the Alps Mountains ranges), and Middle East oil traps (Iraq, Iran, Arabian Gulf region) are all found to be in association with the Alpine-generated structures. The rest of the world major oil fields (in South-Eastern Asia, North and South America) are similarly associated with the Alpine movement.

In association with the Atlas Mountains in North Africa, there are the Algerian and Libyan oil fields, and in association with the European Alps Mountains are the Romanian and Black Sea oil fields. The Middle East area (Iraq, Iran, Kuwait, Saudi Arabia, UAE, Qatar, and Oman), followed by oil provinces in North India, Burma, Indonesia, and Philippine, belong to the Tertiary Alpide belt. In North America, the oil fields are located alongside the Rocky Mountains range, starting from Alberta in Canada to California, Mexico, and continuing alongside the Andes mountain range in South America. Rich oil fields (as Venezuela, Columbia, and Ecuador) are found in the area which is bounded by the Andes Mountains to the West and the Brazilian Shield in the East.

Considered as "oil belts" the main Alpine-related oil provinces (Tertiary age) can be summarized as follows:

- North Africa, Atlas Mountain ranges (Algerian and Libyan oil fields)
- Europe, Alps Mountain ranges (Rumanian oil fields)
- Eastern Europe and Russia, Ural Mountain ranges (Russian oil fields)
- Taurus and Zagros mountains bounding Iraq, Arabian Gulf region oil fields
- South Eastern Asia (Burma, Indonesia, Philippine)
- N. America, Appalachian & Rocky Mountains (USA, Mexico, Caribbean Gulf)
- South America, Indies Mountains (Venezuela, Columbia, Brazil)
- African Rift Zone (Nigeria, Chad, Sudan)

These oil belts are sketched over the world map (Fig. 4.13).

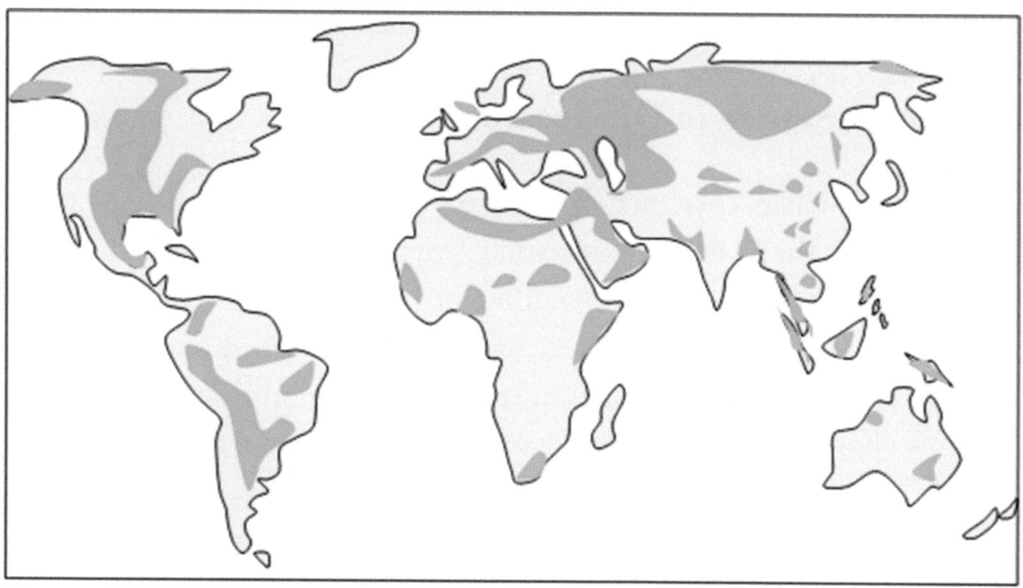

Figure 4.13 Sketch map of the world oil belts

This is a significant map as it shows the genetic link between oil accumulation zones in the world and the global geology. The striking observation is the close association of the major oil zones with the global Alpine tectonic features. Establishment of a system for the distribution of oil in the world (as the oil belt) helps oil explorationists in their prediction efforts.

4.5.3 Global Distribution of Oil Accumulations

Exploration geologists have shown that oil reservoirs are found in sedimentary rocks of various geological ages; from Palaeozoic rocks to the late Tertiary. It is stated that 88% of the World oil is found in the Tertiary and Mesozoic eras. This ratio decreases with increase of geological age, reaching to less than 1% in the Cambrian reservoirs (Assayyab, 1975, p. 263-264). This means that, the majority of the world oil reservoirs are of Mesozoic-Tertiary age, trapped in sandstone-limestone reservoir rocks. Distribution statistics of the discovered world oil are given in the following table (Selley, 1983, p. 4-5).

1. By geological age	
Tertiary	39%
Mezoic	54%
Palaeozoic	7%

2. By reservoir lithology	
Sandstone	49%
Carbonate	45%
Others	6%

3. By type of trap	
Anticline	75%
Fault	1%
Salt dome	2%
Reef	6%
Pinchout	5%
Truncation	2%
Combination	9%

4. By depth (in feet)	
< 4000	10%
4000 –8000	66%
8000 – 12000	22%
> 12000	2%

4.5.4 Terminology System in Oil Accumulations

In the one sedimentary basin, there may be several oil provinces, each of which may include several oil fields. Further, the oil field may consist of several oil reservoirs of different thicknesses and different types of traps and found at different depths. The oil body in a reservoir is sometimes called (oil pool), while an oil-bearing formation is called a pay zone (Fig. 4.14A). The terminology system, normally applied in referring to the oil-bearing zones, is shown in the following flow chart (Fig. 4.14A), and in a sketch of 3D model (Fig. 4.14B).

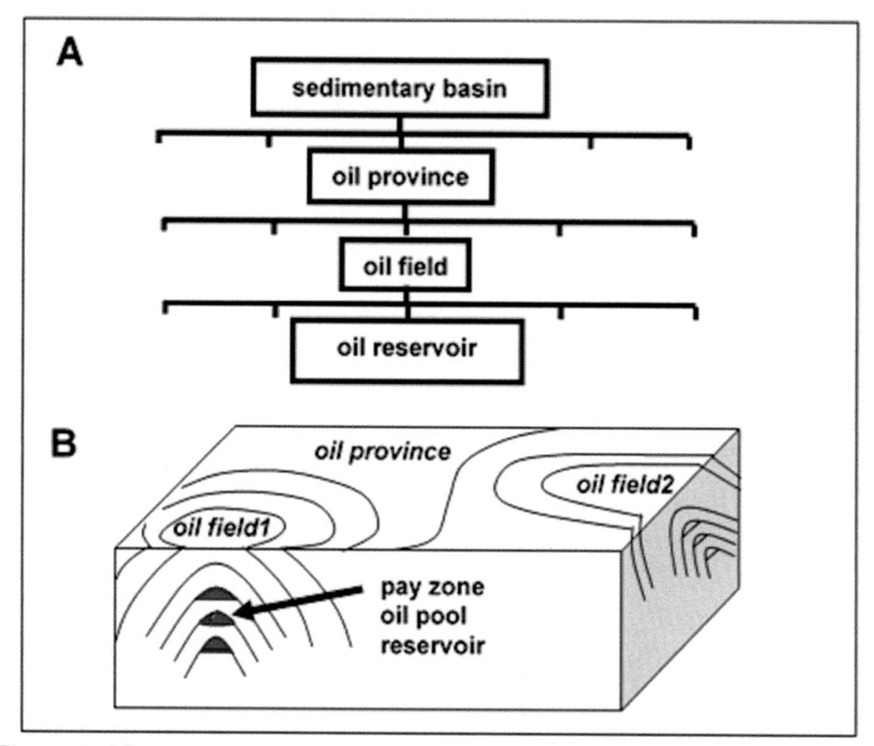

Figure 4.14 Terminology system used in oil and gas accumulation zones. A: flow-chart representation, and B: sketch of a 3D model of an oil province containing two oilfields. Oil field-1 can produce from three pay zones.

Petroleum engineers define an (oil reserve) as the amount of crude oil that can be technically recovered at a financially feasible cost of recovery. When recovery cost is disregarded, the recoverable oil quantities are referred to as (oil resources). One talks about the reserves for an oil well, an oil reservoir, an oil field, a country, a group of countries, or for the whole world. The amount of all found oil (recoverable and non recoverable) in an oil reservoir is referred to as (oil in place).

In this context, it is useful to bring attention to a term that is extensively used in petroleum accumulations discussions. This is the term (petroleum play), or just (play). A play is defined as a group of proved oil-fields or possible oil-fields (prospects), located in the same region and having common geologic environments. The term may be used to refer to the reservoir type of rocks (e.g. carbonate play), to the geological time interval (e.g. Jurassic sand play). Or to a particular geographic area (e.g. Niger delta oil play, southern North Sea gas play). Examples of oil and gas plays are given in (Stoneley, 1995, p106-107).

To summarize, an exploration effort (geological, geophysical, geochemical, …) may provide evidences suggesting possible hydrocarbon trapping structures. These evidences (called leads) are considered to be precursors to (prospects) which are locations where an exploratory well could be drilled to find hydrocarbon deposits. The individual oil accumulation is called (oil pool), (pay zone), or (oil reservoir).

4.6 Oil Reserves

By definition the oil reserves of an oil field, say, is that portion of the oil in place that can be recovered to surface. The ratio of the reserves to the original oil in place is called the (recovery factor). In this respects, and due to its economical importance, world countries spent all possible efforts to explore and determine their individual potential oil reserves. At the end of the day each country now has an officially declared figure for its oil reserve. A common feature shared by all these reserves, and due to technical limitations, oil reserves suffer from uncertainty in the announced estimates. With the general technical advances computation, accuracy is naturally increasing.

4.6.1 Oil Reserves Computation

The next step following discovery of an oil reservoir, is to determine the quantity of the discovered oil. In order to be able to draw a development plan for exploiting the found oil, computation procedure is normally followed to calculate the oil-in-place quantity and its recoverable portion, normally referred to as (the oil reserve). The basic concept underlying the computation is calculating the volume of the reservoir rocks, then the volume of the oil matter which can be produced by the natural dominant reservoir pressure. In its simple form, the oil reserve is computable from the following equation:

OIL RESERVE = VØR(1-S)/F

Where:

V : rock volume

Ø : rock porosity

R : recovery factor (percentage of oil recoverable by the reservoir natural pressure)

S : water saturation (percentage of water mixed with the oil)

F : formation volume factor (volume-percentage of oil shrinkage after production)

4.6.2 Change of Reserve Estimate with Time

Considering the oil reserve of a country, the common observation is that the reserve estimate changes with time, knowing that the oil in place is, by definition, constant quantity. Despite the depletion process, due to continuous production, reserves usually follow an increase or growth trend.

Oil reserves are dependant on exploration effort, which is, in turn, depending on financial budgets available. For this reason, it is observed that the curve, representing its computed value with time, is not uniform, but having abrupt changes. Actual examples for the reserves changes are shown in (Fig. 4.15).

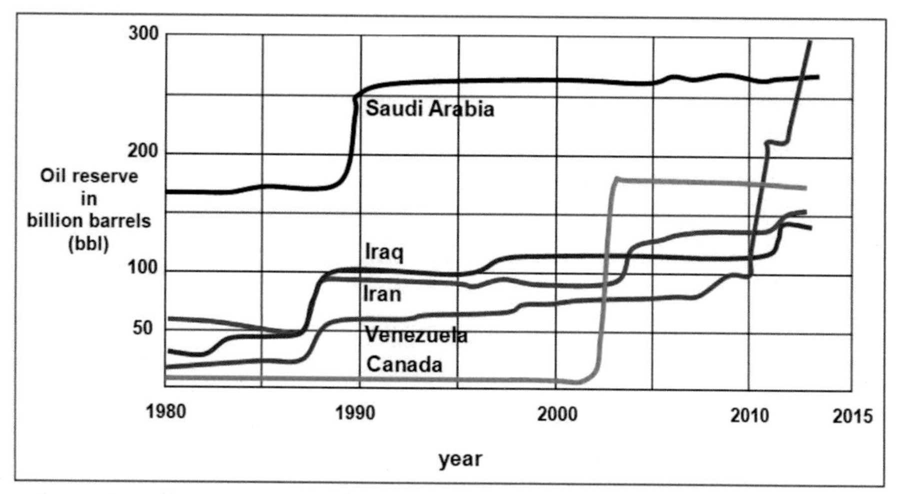

Figure 4.15 Changes of the oil reserves for the top five countries, 1980-2013 (based on data from U.S. Energy Information Administration).

This figure shows the oil reserves changes for the present (2018) top five countries, namely Venezuela, Saudi Arabia, Iran, Iraq, and Canada. The figure indicates that all these countries, with the exception of Canada, have announced a jump in their respective reserves. One of the main factors that brought about the reserve growth is the advances that took place in the exploration technologies which occurred at about in the years of the 1970-1980. Another feature the figure shows is the dramatic jump of the oil proven reserves of Canada which announced in 2003 that oil can be recovered from the oil sands that it possesses by economically rewarding means. Likewise Venezuela declared in early 2011 a sudden increase in its reserves to surpass Saudi Arabia which held the world-highest position up to the year 2011. Such changes in reserves are expected to continue in future times. Considering conventional crude petroleum, and in view of the available exploration data, it is expected that a more dramatic increase in the Iraqi oil reserves will occur in future. Additional extensive exploration efforts can place Iraq at the top of the list of the oil-reserves countries.

4.6.3 World Oil-Reserves

Because of the continuous exploration and production activities (and for other reasons), the declared reserves keep changing from time to time. For this reason published estimates are updated in accordance with the new reserves data. World oil-reserves countries are shown in the following world map (Fig. 4.16).

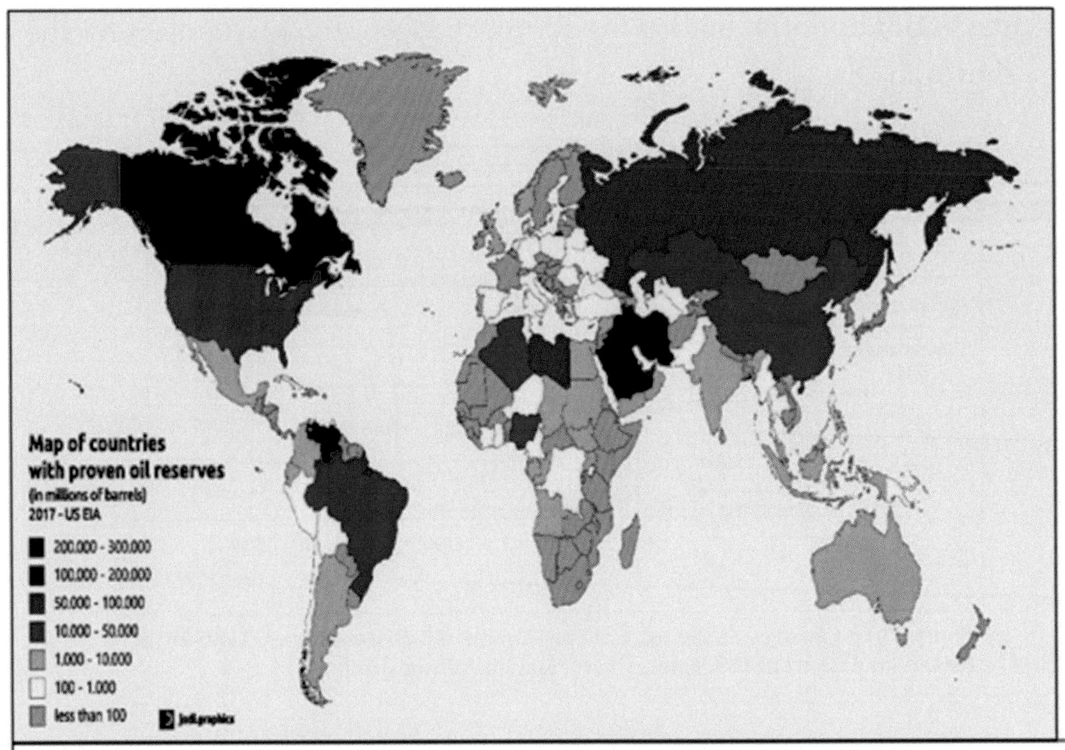

Figure 4.16 World map of countries with proven oil-reserves. Reserve estimates (in million of barrels) are shown here color-coded (U.S. Energy International Administration-2017).

Excluding oil shale sources that boosted the Canadian reserves, this map shows that countries having highest oil reserves are the Middle East countries and Venezuela. In fact, the total oil reserves for the seven countries (Saudi Arabia, Iran, Iraq, Kuwait, UAE, Qatar, Oman and Venezuela) form nearly two third of the whole world's reserves. The following is a table of oil reserves in 2016 for the top 50 countries.

No	country	billion barrel
1	Venezuela	302.250
2	Saudi Arabia	266.208
3	Iran	157.200
4	Iraq	148.766
5	Kuwait	101.500
6	United Arab Emirates	97.800
7	Russia	80.000
8	Libya	48.363
9	Nigeria	37.453
10	United States	32.318
11	Kazakhstan	30.000
12	China	25.620

No	country	billion barrel
13	Qatar	25.244
14	Brazil	12.910
15	Algeria	12.200
16	Mexico	9.711
17	Angola	9.523
18	Ecuador	8.273
19	Azerbaijan	7.000
20	Norway	6.610
21	India	5.749
22	Oman	5.373
23	Sudan & South Sudan	5.000
24	Vietnam	4.400
25	Egypt	4.400
26	Australia	3.985
27	Canada	3.900
28	Malaysia	3.600
29	Indonesia	3.230
30	Yemen (*)	3.000
31	United Kingdom	2.564
32	Syria	2.500
33	Uganda (*)	2.500
34	Argentina	2.185
35	Colombia	2.002
36	Gabon	2.000
37	Congo, Brazzaville (*)	1.600
38	Chad (*)	1.500
39	Brunei	1.100
40	Equatorial Guinea (*)	1.100
41	Ghana	0.660
42	Turkmenistan	0.600
43	Romania (*)	0.600
44	Uzbekistan	0.594
45	Italy (*)	0.557
46	Peru (*)	0.473
47	Denmark	0.491
48	Tunisia (*)	0.425
49	Thailand (*)	0.396
50	Ukraine	0.395

Table of the oil reserves in 2016, for the top 50 countries (OPEC Annual Statistical Bulletin-2017). Data marked (*) were obtained from the U.S. Energy International Administration (EIA)-start of 2017.

A point needs clarification and that is the oil reserves of Canada. The EIA quoted the figure (169.709 bbl) based on the 2003-Canadian announcement of additional oil reserves obtainable from oil sands. According to (OPEC Annual Statistical Bulletin-2017) the world total oil reserves in 2016 is 1492.164 bbl

4.6.4 World Gas-Reserves

Natural gas is the gaseous hydrocarbon which consists primarily of methane and ethane compounds. It is commonly found as part of an oil reservoir (oil field) or as an independent gas body (gas field). Depending on the pressure conditions, the gas may be found as dissolved in the reservoir oil or as a separate free gas (gas cap) accumulated over the oil zone. The other type of gas source is what is termed (tight gas) which is the gas found trapped in reservoir rocks having low porosity and low permeability. Tight gas reservoir rocks may be shale, sandstone, or limestone. Special drilling programs and adequate production technology are developed to achieve economically rewarding gas-production.

As it is with the case of oil reserves, gas-reserves of countries change with time. The principal factor for the jumps witnessed by the country gas reserves is the technological advances that made gas producing from tight reservoir rocks, economically feasible.

No	country	trillion cubic meter (tcm)
1	Russia	50.485
2	Iran	33.721
3	Qatar	24.073
4	United States	9.184
5	Saudi Arabia	8.619
6	Turkmenistan	9.870
7	United Arab Emirates	6.091
8	Venezuela	5.740
9	Nigeria	5.475
10	China	3.611
11	Algeria	4.504
12	Iraq	3.820
13	Australia	3.205
14	Indonesia	2.775

No	country	trillion cubic meter (tcm)
15	Malaysia	2.740
16	Kazakhstan	1.907
17	Egypt	2.086
18	Canada	2.181
19	Norway	2.362
20	Azerbaijan	1.284
21	Uzbekistan	1.585
22	Kuwait	1.784
23	Libya	1.505
24	India	1.458
25	Ukraine	0.952
26	Oman	0.885
27	Netherlands	0.799
28	Pakistan	0.723
29	Brazil	0.392
30	Bangladesh	0.385
30	Peru	0.384
31	United Kingdom	0.336
32	Angola	0.308
33	Bolivia	0.301
34	Trinidad & Tobago	0.290
35	Myanmar	0.282
36	Mexico	0.270
37	Brunei	0.258
38	Thailand	0.220
39	Vietnam	0.207
40	Cameroon	0.153
41	Congo	0.115
42	Colombia	0.113
43	Romania	0.103
44	Denmark	0.072
45	Poland	0.061
46	Italy	0.047
47	Germany	0.046
48	Chile	0.040
49	Gabon	0.026
50	Ecuador	0.011

Table of the gas reserves in 2016, for the top 50 countries and the world total is 200.539 tcm (OPEC Annual Statistical Bulletin-2017.

This table shows that the gas reserves are dominated by Russia, Iran, and Qatar. These three countries together hold about half the world's proven gas reserves.

4.7 Unconventional Oil and Gas

By nature, energy types such as the solar energy, wind-energy, and hydro-energy are typical forms of renewable energy. All types of energy sources (conventional and renewable) are providing the necessary energy for electricity generation, heating/cooling processes, transportation activities, and industrial operations.

Energy sources provided by biological remains which were preserved in the Earth upper Crust are termed (fossil fuels) which include natural gas, oil, and coal. With the world-wide production, these resources are continuously depleted, and because they are not liable to rejuvenation, they are not considered as renewable type of energy sources.

4.7.1 Basic Definitions

In conventional work, oil is withdrawn from its habitat oil-trap, through drilled oil wells. In many parts of the world, there deposits of hydrocarbon matter found in rock formations in the form of high-viscosity oil, tar, or with immature hydrocarbon (Kerogen), mixed with the rock matter. The most commonly known rock types which have hydrocarbon matter found as part of the rock texture are sandstone and shale formations.

The unconventional hydrocarbons are the oil and natural gas (including the gas obtained from coal) which are produced from unconventional hydrocarbon sources using methods other than normal oil-well production. In this case the hydrocarbon matter is tar or highly viscous oil mixed with the rock particles of the formation and the means of extraction is through normal mining or by hydro-fracturing techniques.

4.7.2 Types of Hydrocarbon Resources

A hydrocarbon resource is defined to be the total amount of hydrocarbon deposits present in a certain location, regardless of whether the deposits can be extracted or not. Oil and natural gas produced from normal hydrocarbon traps (by the known normal well-drilling technologies) are commonly known as the conventional hydrocarbon resources. In nature, there exist other types of hydrocarbon deposits made up of high viscosity oil or bitumen mixed with the rock texture of sedimentary rock-formations (usually shale or sandstone). This type of hydrocarbon, from which oil and gas can be extracted, is normally referred to as unconventional hydrocarbon resource.

Commonly known unconventional hydrocarbons come from two main types of rocks; sandstone formations (called tar sands) and shale formations (the oil shale). A third example of the unconventional hydrocarbons is natural gas found associated with coal. The hydrocarbons (oil and/or gas) extracted from tar sands, from oil shale, or from coal are typical forms of unconventional hydrocarbons.

4.7.3 Tar Sands Hydrocarbon Resources

Tar, as a substance, was known in ancient Iraq some 5000 years ago. It was used by the Sumerians and Babylonians in their municipality works (construction of buildings, roads, water canals, and boats). At present there exists a large tar-deposit resource outcropping on surface, near Hit city, located some 100 km WNW Baghdad, Iraq.

Tar sands or oil sands, as it is also called, are mixture of sand, clay, water, and heavy hydrocarbon matter (dark and highly viscous oil) having the nature of tar. In its natural state, the hydrocarbon part of the tar sands cannot be pumped to surface as in the case of conventional oil production.

4.7.4 Oil Extraction from Tar Sands

To extract the contained oils, tar sand deposits have to be mined using open pit or other types of mining techniques. Other ways and means applied to extract the oil is done by underground heating with additional oil-isolation processes. Before sending the so-obtained hydrocarbon to refineries, special processing is needed to be carried out. This means that the unconventional hydrocarbon resources require complex production and processing technologies which means high financial cost.

4.7.5 The Canadian Tar Sands

Tar sands exist in many countries around the world. In particular, the four countries (Canada, Venezuela, Russia, and Kazakhstan) have exceptionally huge resources. Canada. Recent reports stated that the Canadian tar-sand resource is estimated to be over 2 trillion barrels. The main Canadian deposit prospect is the (Athabasca Tar Sands), found in Alberta province. Geologically, the Athabasca tar-sand layer is a lower Cretaceous formation, called (McMurray) formation.

According to the U.S. Energy Information Administration (2015), the total Canadian proven oil reserve, including recoverable tar-sands oil, is estimated to be 172 billion barrels. This figure places Canada's oil reserves in the third place (after Venezuela and Saudi Arabia) among the world top oil-reserves countries.

4.7.6 Oil Shale Hydrocarbon Resources

Oil shale is a fine grained sedimentary rocks containing relatively large amount of immature hydrocarbons (Kerogen), from which significant amounts of normal oil and hydrocarbon-gas, can be extracted. Specially-developed industrial techniques were developed to extract oil and gas from oil shale sediments.. Geologically, oil shale were formed under continental, marine, or lacustrine depositional environments and they can be of any geological age, from Cambrian to Tertiary.

4.7.7 Oil Extraction from Oil Shale

Methods applied in extracting oil from oil shale are similar to those used in the case of tar sands processing. Shale oil is extracted by pyrolysis (heating in an inert atmosphere) or by thermal dissolution processes. At present shale-oil extraction processes are mostly done on the oil-shale rocks after being mined. In comparison with tar sands, oil shale extraction-processing is reported to be even more pollutant to environments.

4.7.8 The USA Oil Shale

Like tar sands, oil shale deposits are found in many countries, but the largest known deposit is found in the Green River formation (Eocene age) in western United States. According to the World Energy Council, the United States shale-oil resource in 2016 could reach 3.7 trillion barrels.

According to the report (World Energy Resources-2016) issued by the World Energy Council, the total world resources of oil shale is equivalent to yield of 6.05 trillion barrels of oil. Considering the present world total conventional oil reserves to be at 1.5 trillion barrels, it can be concluded that the global shale-oil resources amount to 4 times as large as the world total conventional oil resources.

4.7.9 Coal-Bed Gas

The coal matter of a coal bed is a very low permeability, but it has naturally developed system of fractures (called cleats). This fracture system is made up of two perpendicular sets of fractures which are both at right angles to the coal bedding. The gas (mainly methane gas), associated with coal is a found accumulated within the spacing of the cleats fractures.

The United States is estimated to have 270 Tcf of recoverable coal-bed gas (Hyne, 2012, p. 479).

4.7.10 Effects of Unconventional Hydrocarbons on Environments

Though the unconventional hydrocarbon resources are so abundant, their exploitation is handicapped by production high-cost and complex technologies, in addition to the serious pollution effects inflected to environment. Effects on environments come from all those activities associated with oil and gas, on extraction from the raw oil sands and oil shale deposits. In particular, there are the processes ; open-pit mining, atmospheric carbon dioxide emissions, disposed waste material, and pollution of the over-ground and under-ground water.

The incurred damage, resulting from unconventional hydrocarbon production industry is found to be far more severe than that resulting from conventional-fuels exploitations. The environmental impact of all activities concerning exploitation of the unconventional fuels is now becoming a serious public issue.

Chapter-5

5. OIL EXPLORATION TECHNIQUES

As we have learnt from the previous chapter hydrocarbon accumulations are normally found within the pores of subsurface sedimentary rocks of suitable geometrical shapes; the oil traps. Realization of the whole set of elements of the petroleum system is essential prerequisites for creating an oil or gas body entrapped in a subsurface oil trap. Since oil traps are normally formed at depths, exploration techniques need to be applied in order to locate them and to get as much knowledge as possible about their nature.

5.1 The Oil Exploration Work

The conventional oil-exploration methods (usually by seismic techniques) are considered to be indirect means, in the sense that they aim at finding geological structures capable of trapping hydrocarbons. These structures may, or may not, contain oil. The other type of exploration methods (called direct hydrocarbon detection, DHD methods) are applied with the objective of detection of hydrocarbons if they actually exist. However, both of the direct and indirect exploration approaches are always involving some degree of uncertainty. Exploration risk can be reduced with availability of additional geological information.

5.1.1 Oil Exploration Main Objectives

The principal objective of any oil-exploration project is to locate subsurface geological structures capable of holding hydrocarbon matter in commercially-rewarding quantities. Exploration activities involve utilization of a number of geological and geophysical techniques which are applied with the most optimum parameters with which exploration objectives can be achieved.

Usually oil exploration activities involve the following objectives:

- Identification of oil traps, their types, depths, and dimensions.
- Knowing oil-reservoir statistics, oil quantities and types of fluids.

- Deduction of the geological and reservoir prevailing conditions.
- Computing the recoverable oil-reserves.

5.1.2 The Exploration Effort Evaluation

Oil exploration is, in fact, a continuous process which does not end with finding oil, but continues after drilling the first exploration-well. This is a necessary procedure in gathering the data required in drawing the field-development plans for the evaluation and development drilling programs. The ratio between the cost of total exploration effort and that of the discovered oil represents an important parameter that provides an expression for the efficiency of the executed exploration process.

Normally, when an oil field starts production, its oil content gets depleted and its reserve is gradually reduced. With no additional discovery to make up for the depleted oil, the exploration effort parameter gets lowered. Thus, in order to keep this parameter at its original level, exploration activities need to be done to bring about additional new oil discoveries or to bring about conversion of probable reserves into proven reserves.

5.1.3 Exploration Work Strategy

Basically, an oil explorationist, in his oil exploration work, is equipped with the well-established principles concerning oil generation, migration, and accumulation. In other words, exploration work-plans are guided by the conditions upon which the petroleum system of the target area is based. This means that the exploration work strategy involves finding not only oil traps but all of the elements of the petroleum system for the area under study. Exploration efforts shall aim at finding depths and extensions of the reservoir rocks and the source rocks capable of generation of mature hydrocarbon matter. The study involves identification of the oil migration path ways. It is important also to determine the geological and geophysical properties of the involved rock medium.

An important item of the oil exploration is determination of the location of the target area in relation to the sedimentary basin (or with respect to regional tectonic features) surrounding that area. Source rocks may, by the way, be located within, or outside, the assigned exploration area.

Determination of the exploration area in relation to the basin centre is an important issue since this will help in assessment of the sedimentary model of the subsurface geology and, hence, prediction of related geological features as the basement depths, facies changes, and effecting tectonic changes. The study should lead to a general geological model that shows the principal elements of the existing petroleum system which include the source rocks, reservoir rocks, the cap rocks, and the trapping mechanism.

5.1.4 Trends of Geological Studies

In order to produce a subsurface geological model of the exploration-area, two main trends of investigations are required to be followed. These are the structural and the stratigraphic studies.

(i) The Structural Studies.
Part of the exploration investigations is to establish the tectonic and structural pictures of the area and to determine specifications and historical developments of the folding and faulting features. Establishing scenarios of the large-scale tectonic features and the detailed structural changes which have taken place at the time of, and after the development of the sedimentary basin is important in helping of tracking possible migration routes and in visualizing expected oil traps.

(ii) The Stratigraphic Studies.
For more complete picture of the subsurface geology, studies need to cover the stratigraphic changes taking place within the sediments of the sedimentary basin. As in the case of the structural features, stratigraphic features help in identifying stratigraphic and structural-stratigraphic combined traps.

Generally speaking, exploration of the stratigraphic traps is more difficult than those of structural types. This is because of the relatively weaker geophysical anomaly compared with those associated with the structural changes. The main elements of the stratigraphic study are:

- Formation lithology, extensions, and geological age.
- Sedimentary environments and facies distributions.
- Petrophysical properties and reservoir characterization.

5.1.5 Summary of Exploration Work Strategy

In order to produce a clear geological model of the exploration-area, the exploration geologist presents geological profiles through the basin model. Other useful aiding drawings are also made. Examples of such aiding techniques are constructing regional structural maps, isopach maps for certain selected formations, and palaeogeographic maps. These are of great help in predicting the sedimentation sources, sedimentation routes, and their deposition zones.

Exploration-work strategy can be summarized in the following sequence of steps:

1	regional tectonic geology
2	local structural geology
3	local stratigraphic geology
4	sedimentary basin outlines
5	petroleum system elements
6	petroleum potentialities and reserve estimation

The investigation work strategy is implemented in steps starting with covering the large-scale regional geology then into detailing geological and reservoir characterization. For each of these steps the collected data can be obtained from available geological, geophysical, and petrophysical data, obtained from wells drilled inside, or outside, the exploration-area. Another more recent source of information used in geological studies are satellite imagery and aerial photography. These techniques have proved to be effective additional tools used in exploration studies.

5.1.6 Nature of Oil Exploration Science

Oil exploration is considered to be an inexact science, or science of speculative nature, sine it is based on a number of speculations and prediction approaches. The other characteristic nature of oil exploration practices is that it is based on practical observations and actual field measurements. The gathered observation data is then subjected to data refining (data processing) followed by interpretation work that leads to determination of the subsurface geological model.

Effectively, the total exploration process consists of a set of three main steps applied in series. These steps are: collection of the observational data (field work), refining computations (data processing), and deduction of the subsurface geological structure (interpretation of the processed data). In view of the limitations which these activities (incomplete data and statistical approaches), oil exploration becomes an inexact science having a statistical nature.

It is interesting to note that there is a great resemblance between the work of the oil explorationist and that of the medical doctor. Both of the geological exploration and illness diagnosis are, in a sense, considered to be inexact sciences. The exploration geologist needs to be equipped with the knowledge of the anatomy of the Earth crust and with the tectonic and stratigraphic activities just like the medical doctor who, likewise, needs to have full understanding of the anatomy of the human body and functioning of the body various organs. In the medical field, diagnosis is made, based on observations and measurements of the changes which occur as a result of the illness or any other mishap that happen to the patient. In the same way, the exploration geophysicist carries out observations and measurements of the Earth natural phenomena (such as gravity and magnetic fields), then comes out with the appropriate diagnosis of the causing geological anomalies.

5.1.7 The Oil Exploration Project

In spite of the statistical nature of the exploration outcomes, oil exploration projects are considered to be in the rank of the large industrial projects. Oil exploration projects have the following characteristics:

- It involves great financial investments.
- It involves great degree of economic risk.
- It requires multi scientific and engineering specializations.
- It depends on technologies which are rapidly changing.
- It depends on modern automation, electronic devices, and IT techniques.
- It needs large number of highly qualified and trained personnel.

-Its findings may inflict strategic impacts on states and peoples, leaving marked imprints on international political and economical relationships. It can lead to political conflicts and it may even lead to waging wars.

5.2 Geological Exploration and Aerial Photographs

Naturally the geological environments are responsible to formulations of the elements of the petroleum system. The close relationship between geology and the elements of the petroleum system expresses the important role of geological knowledge in reaching successful prediction of the source rocks and potential oil traps.

5.2.1 Surface Geological Surveying

Though it is important, conventional surface geological surveying is now rarely applied in the exploration process of oil. However, this method involves examinations of rock outcrops (measurements of dips, strikes, and thicknesses). The studies also include identifications of fossils and diagnoses of lithological types, and mineral contents. By scouting activities, oil seepages may be discovered and evaluated.

Geological data are gathered from surface outcrops as well as from wells, previously- drilled in the area. These data can be presented in form of geological profiles and sections. The gathered data may be sufficient in showing the three dimensional picture of the geology of the area, showing the stratigraphic units and geologically-identified rock formations.

In addition to profiles, topographic and geological contour maps can also be constructed. Surface topographic features, drainage system, and changes of vegetation types can be used by the geologist as indicators of subsurface geological features.

5.2.2 Geological Correlation Technique

A powerful geological tool which is often made in this connection is drawing correlation charts. By an interpolation approach (called geological correlation) is the technique used in determination of continuous formations. The main criteria used in the correlation process are geological age, lithology, and fossil contents. The method is shown in (Fig. 5.1).

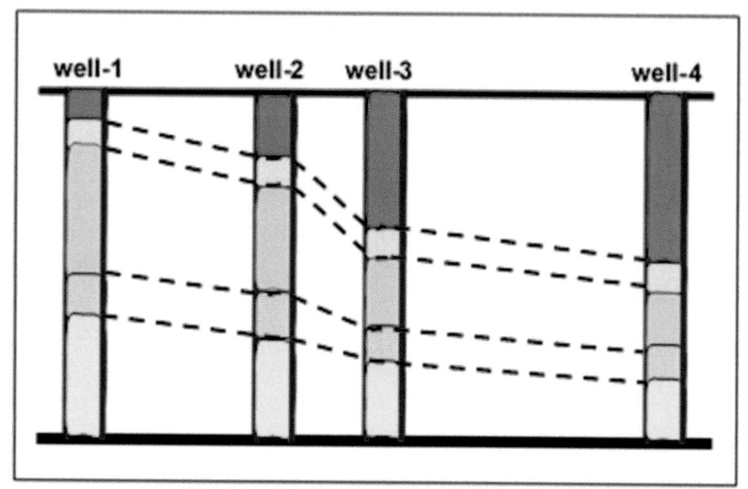

Figure 5.1 The geological correlation process

From the palaeontological studies of the rock formations penetrated by the drilled holes, the geological age of the formations are determined. In practice, such studies are relying more on the microfossils (as foraminifera, botanic pollens and spores) which belong to a recently-developed branch of science; Palynology. This type of studies proved very effective in the correlation processes. Palynological studies are normally carried out by use of modern electronic equipments like the electron microscope, while geological ages are determined (by radioactive dating method) from the decay-rate of the radioactive minerals found within the rock formations.

The correlation method gives more accurate results, the closer the observation points (well sites) are, and in cases where the wells are all belonging to the same sedimentary basins and not distributed among several separate basins.

5.2.3 Construction of Palaeogeographical Maps

In addition to the stratigraphic correlation tool, the exploration geologists can draw palaeogeographical maps which show the geographical features at certain geological ages. These maps are helpful in showing sedimentation source areas, related deposition routes, and types of deposition environments (as fluvial, deltaic, lacustrine, marine). These maps, together with the regional structural and stratigraphic maps, serve as guides for determining the elements of the petroleum system of the area under investigation.

5.2.4 Aircraft and Satellite Photographing

One of the relatively recent exploration-techniques is the use of aerial photographs produced by a special technique; the remote sensing technology. Two types of such activities are in common use. These are aircraft photography and satellite imagery, briefly defined as follows:

- Aerial Photographs

These photographs, taken from airplanes or from satellites, give direct pictures (photographs) of the surface changes including geological, geographical, and vegetation changes. These photographs are subjected to interpretation for purpose of deriving the geological changes. The science specialized in this type of interpretation is called (Photogeology). The main types of information obtainable from the aerial photographs are: topography changes, rock-outcrops, drainage system, and vegetation, in addition to geological structural and stratigraphic changes.

- Satellite Imagery

Satellite-mounted radar equipments are employed to emit high-frequency electromagnetic signals and record these signals after being reflected from the Earth surface. These records are then processed and presented to the interpreter in the form of color-coded images. This type of exploration, which are depending on radio waves generated by electronic devices mounted on aircrafts flying at high altitudes or from artificial earth satellites, is usually referred to as (remote sensing).

Personnel, specialized in interpreting aerial photographs and satellite imageries, give their interpretation results showing geological anomalies and oil seepages, if they are there.

5.3 Geochemical Exploration

Surface geochemical surveying depends on direct analyses of hydrocarbon matter that has seeped through from subsurface oil accumulations. Historically, searching for oil seepages is considered to be the first direct geochemical method applied to find oil. In fact the first oil well, drilled by Drake in 1859, was located on the basis of oil seepage appearance.

An oil seepage occurs as a result of dissipation of hydrocarbon matter (oil or gas) through channels connecting oil accumulations with the Earth free surface. The resulting seepage may be large-enough, readily seen (macro-seeps) or very small (micro-seeps) which are too small to be visually identified. In this case (case of micro-seeps), it can be ascertained either by using special hydrocarbon-detectors or by interpretation of the side effects which are incurred to the vegetations growing in the affected area.

Usually, geochemical data are supported by the remote-sensing images, provided by airborne photographing equipments and by satellite imagery data.

5.3.1 Geochemical Methods

There are two main geochemical approaches which can be followed in surface exploration of hydrocarbons.

A direct method can be used for detection of traces of hydrocarbon compounds (as hydrocarbon gases) which may be present in the surface soil (land geochemical surveying) or in the sea water (marine geochemical surveying). The other approach (indirect method) is depending on the side effects which are caused by hydrocarbons on the area. Due to presence of hydrocarbon traces, chemical, physical, and micro biological effects (on surface vegetation) may be observed.

In the direct method, a uniform grid of lines is drawn for the area, where the cell dimensions are something like (1km x 1km) or (½ km x ½ km). From each grid-point a soil sample, taken from depth of about one meter, is taken and subjected to chemical analysis, usually by use of the (gas chromatography) analyzer. In case of marine geochemical survey, the water sample is taken from depth of about ten meters above the sea floor.

These geochemical methods can, in general, give indications on the existence of subsurface hydrocarbon sources, but they are incapable of determination of the actual locations, dimensions, or depths of these sources.

5.3.2 Geochemical Well Logging

A geochemical log is defined as the record which expresses the variation of degree of concentration of a certain chemical element with depth. Construction of the geochemical log of a radioactive matter can be done with the aid of plotting the variation of natural gamma radiation with depth.

In addition to the logging technique, there is the chemical-analysis which is done for rock samples obtained from the drill-hole during the drilling operation. The analyses are done at well sites or in special office laboratories. The principal purpose of these analyses is determination of types and quantities of organic matter in the rock samples collected from the well during drilling. An important geochemical analysis is the so-called (pyrolysis). This type of analysis is based on heating a rock sample then measuring certain parameters like quantities and types of hydrocarbons found in the collected rock samples. Also, pyrolysis is used to determine percentages of associated chemical elements as sulfur, carbon, Oxygen, and Hydrogen.

5.3.3 Applications of Geochemical Analyses

Most important application of the geochemical analyses is in locating the oil accumulations and in assessing oil properties. In order to achieve these objectives, chemical analyses are carried out on rock samples taken from well cuttings and cores. Such analyses are also done for oil and gas samples taken from hydrocarbon matters found in association with cores and cuttings. The results of the analyses are normally used in evaluation of source and reservoir rocks (as lithology, petrophysical properties, and geological age).

The other important field of application of the chemical analyses is determination of the generation-location, migration-routes, maturation degrees and its variation with depth. The obtained analysis-data can give information on facies changes, unconformities, and geological timing of these activities with respect to precipitation and tectonic activities. Drilling engineers use these data in decisions as regards definitions of abnormal high-pressures and as guides for decision on well test-zones locations.

5.3.4 Evaluation of Source-Rocks

Source rocks are characterized by being rich in organic matter (Kerogen) which has passed sufficient degree of thermal maturation that produced hydrocarbons. The accepted theory of the primary migration of the produced oil is that the generated oil gets expelled to the neighboring porous rocks by the pressure-buildup associated with the oil-generation process caused in the source rocks.

The common approach of evaluation of the maturation degree is by:

- chemical analysis
- thermal analysis (pyrolysis)
- optical examination of the organic matter by use of microscope to identify micro-plants, Palynological remains (spores and pollen).
- Reflectance test, especially for the reflectance of the mineral (Vitrinite) which gives indication of the origin and types of the constituents of the organic material.
- Fluorescence test using ultra violet (UV) light excitation which can be an indicative on presence of organic matter.

These geochemical tests and analyses made for rock cores and rock cuttings from several wells can be used in source-rock correlation and in oil-oil correlation. Concentration and distribution of the chemical elements (N, S, O), are used in the oil-oil correlation process. Gas chromatography is used in studying the properties of bituminous matter and crude oils, including percentages of various types of hydrocarbon series.

A closely related issue is construction of the thermal maturation curve which can be derived from variation of the matured hydrocarbon proportion with depth. From these data, sedimentation and burial history may be inferred. Determination of the geothermal gradient is another approach used to give a model that describes the oil generation and its development history. In general, it is found that the thermal gradient becomes high in the rifting areas and in other areas where folding and where volcanic activities exist. Thermal gradients are found to be moderate in tectonically-inactive places.

5.4 Geophysical Exploration

It so happened, that oil accumulations are often found at depths of few kilometers, and not found at the Earth surface as in case of water which mostly exists in the form of rivers, lakes, and seas. In order to determine locations of the deep oil reservoirs, exploration ways and means have been devised, developed and put in profitable applications. During the past hundred years or so, several effective geophysical techniques, have been developed. In the following discussions these techniques shall be briefly reviewed.

5.4.1 Phases of Geophysical Exploration

A geophysical exploration-survey normally follows a sequence of activities starting with field work and ending up with determination of the subsurface geological model. A typical exploration survey-project normally passes through a sequence of three operation phases.

The starting phase of a survey (acquisition phase) is collecting of the field raw-data through standard field procedures by which the geophysical measured values are recorded, usually on a certain storage medium. After completing the acquisition work, the recorded data are passed on to a processing centre where it is subjected to certain processing steps (processing phase). This is done for purpose of certain corrections and for enhancement of the geophysical signal. The third and last phase of the project (interpretation phase) is to interpret the final processed data to extract the subsurface geological model of the area under exploration. These three phases of exploration work can be summarized as follows:

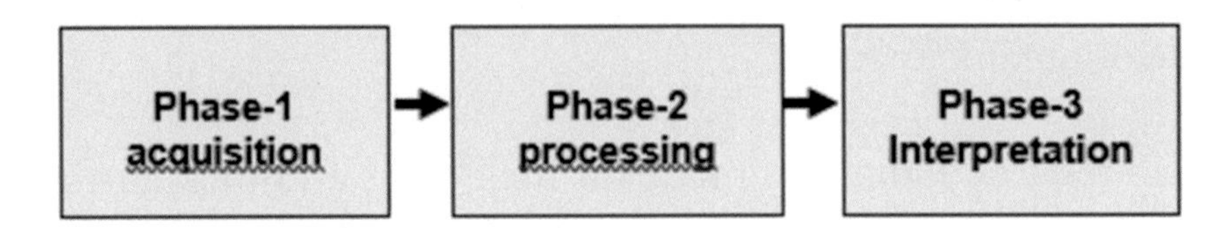

It is apparent from this flowchart that geophysical exploration is an inverse problem in which a geophysical response (measured in the acquisition phase) is, after being processed, interpreted, giving the geological model responsible for that geophysical response.

5.4.2 Geophysical Exploration Methods

It so happened, that oil accumulations are often found at depths of few kilometers, and not found at the Earth surface as in case of water which mostly exists in the form of rivers, lakes, and seas. In order to determine locations of the deep oil reservoirs, exploration ways and means have been devised, developed and put in profitable applications. During the past hundred years or so, several effective geophysical techniques, have been developed.

The most commonly applied techniques are based on use of the Earth potential fields (as gravity and magnetic fields) and seismic waves. Other methods (electrical and electromagnetic fields) are, in practice, not so much in use for oil exploration. Seismic method is by far the most widely applied method in the exploration of oil and gas deposits. The rest of geophysical techniques (Gravity, magnetic, electrical, and radioactivity methods) provide support data in the oil exploration and also applied in well logging. These methods are defined as follows:

5.4.3 The Seismic Method

This method is based on generating seismic waves by a mechanical energy source at a point (S) on or just below the ground surface and recording the arrivals, at other surface points (R), of the reflected waves. By repeating the operation of wave generation at a point and recording the arrivals of the reflected waves, the survey area can be covered. Ray diagram and an actual record of a typical seismic shot point, are shown in the following diagram (Fig. 5.2).

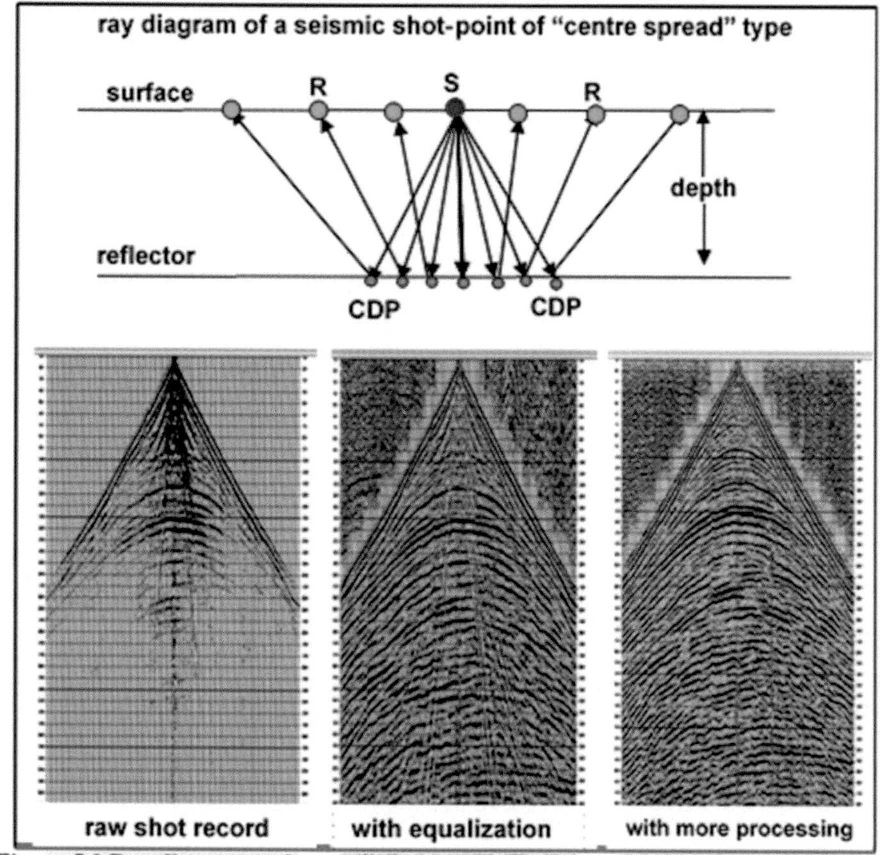

Figure 5.2 Ray diagram and actual record of a typical seismic shot point

From the travel-time measurements of reflection waves and wave motion-velocity, the structural variation of subsurface geological layers are mapped. Under favorable conditions, study of the seismic wavelet can provide information on the stratigraphic nature and hydrocarbon contents of the traversed rock formations. The raw seismic data (normally recorded on magnetic tapes), are passed through a sequence of processing steps followed by a set of certain interpretation procedures in order to obtain the final result, which is the subsurface geological model.

In any seismic experiment, and in general, four types of seismic waves may be generated. These are the direct, refracted, reflected, and the diffracted wave arrivals. Refraction and reflection of waves occur when incident wave hits an interface and the diffracted waves are created in case of existence of point diffractors (Fig. 5.3).

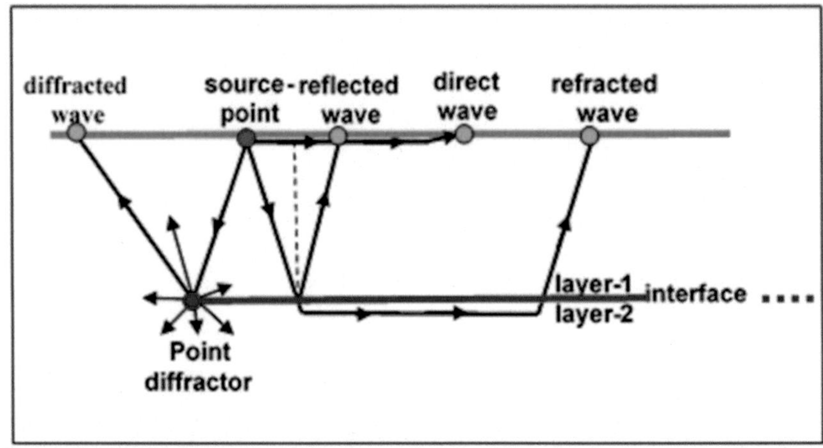

Figure 5.3 Main types of seismic waves, involved in seismic-survey works. From a terminating interface, four wave arrivals can be observed (direct, reflection, refraction, and diffraction).

In the conventional 2D seismic survey, shot records are obtained for points uniformly distributed along two sets of straight lines (seismic lines). One set is laid down in the dip-direction, and the other set is in the strike direction. The outcome of the processing work (involving various types of corrections) is getting a seismic stack section for each of the surveyed seismic lines. The whole process may be summarized as in the following diagram (Fig. 5.4).

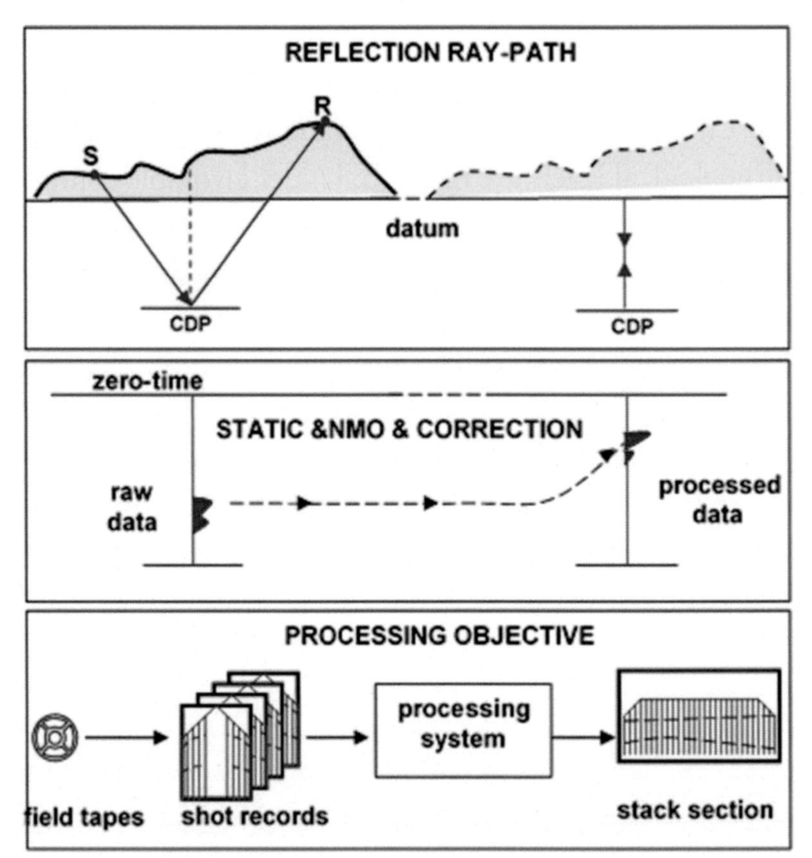

Figure 5.4 Reflection ray-paths, corrections, and the final processing outcome, stack sections from the field shot records.

The seismic reflection method is considered to be the most effective tool applied in oil exploration. In terms of exploration effort, it is rated at 95% of all geophysical techniques used in oil-exploration. The intensive application of the method is due to its well-established potentialities in revealing the subsurface geological structure, down to depths reaching to more than 10 km. Since it became in common use (in early 1930s), the method proved very successful in the discovery of oil traps, especially structural types of traps. Examples of actual seismic stack sections showing folded and faulted strata are shown in (Fig. 5.5).

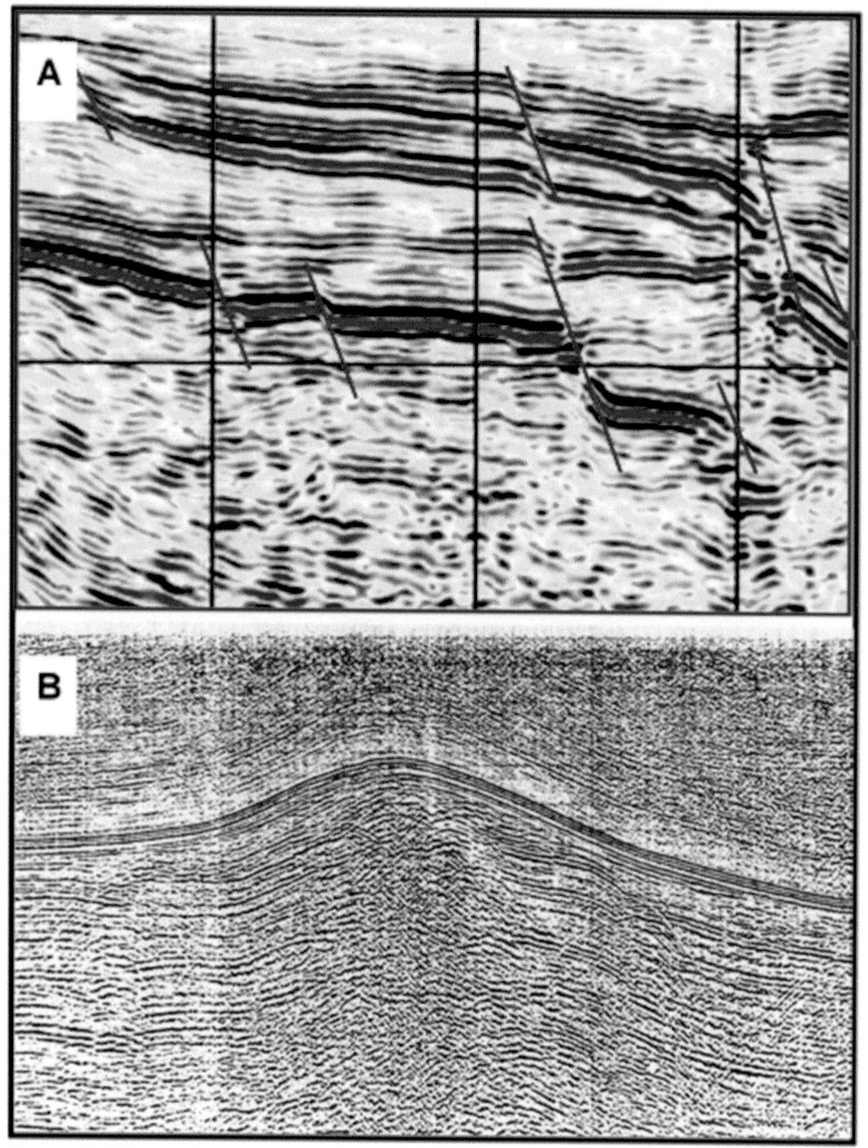

Figure 5.5 Examples of stack sections from seismic reflection survey of an area (A) having faulted- and an another area (B) having folded-structural geology.

Under favorable conditions, the method can give valuable information on stratigraphic changes and on direct hydrocarbon detection. Examples of stratigraphic features revealed by the seismic reflection method are given in (Fig. 5.6).

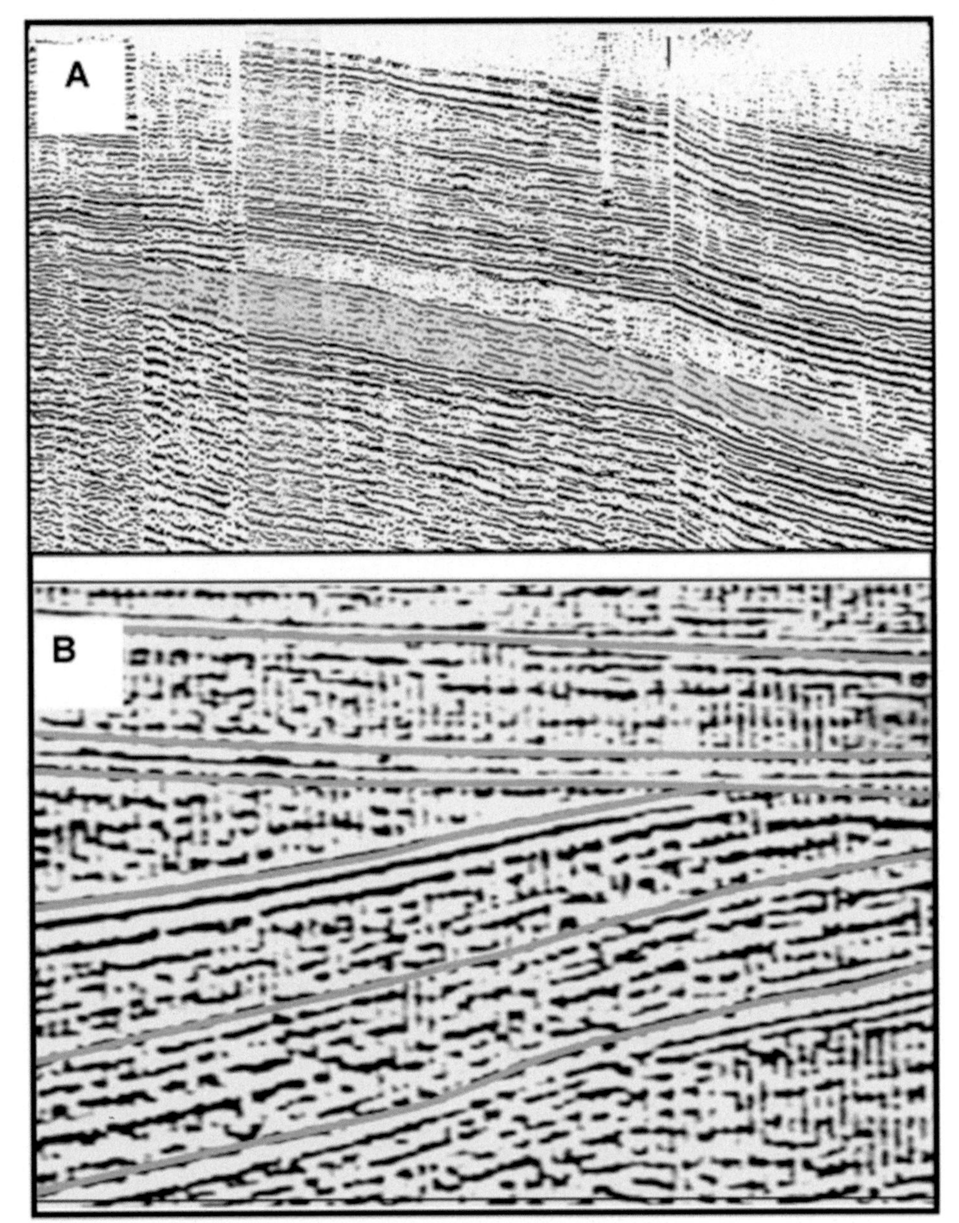

Figure 5.6 Examples of stratigraphic features revealed by the seismic reflection method. (A-Giant sand lens) & (B-angular unconformity)

Since the introduction of the method, more than ninety years ago, the seismic reflection method has witnessed many technical developments. The use of the vibratory seismic source (Vibroseis), the magnetic tape recording, the digital data processing, and use of seismic amplitude for stratigraphic and direct hydrocarbon detection, is prominent features of those developments.

The other important development is the introduction, in the 1970s, of the three dimensional (3D) surveying technology. With this type of surveys, a 3D image (seismic image) is obtained for the whole surveyed area. In this case, horizontal sections, as well as vertical sections, can be obtained from the survey data-box (Fig. 5.7).

Figure 5.7 Examples of a 3D data volume and a horizontal seismic section obtained from processing of a 3D data-set,

All these developments have made the seismic reflection technology by far the most effective oil-exploration tool.

Numerous books have been published on the seismic exploration method which can be referred to for getting detailed information. Examples of such references are: (Dobrin, 1960), (Dobrin, and Savit 1988), (Sheriff & Gildart, 1995), (Alsadi, 1980), and (Alsadi, 2017).

5.4.4 The Gravity Method

Based on Newton's Law governing the gravitational force of attraction, the gravity value (measured in acceleration units) of a buried body-mass, is larger with high density and lower with deeper burial. In general, a geological anomaly (as a body mass of density different from that of the host medium) will create a corresponding gravity anomaly of value which is greater as the density contrast is greater. Further, the gravity anomaly of a deeper rock body is less than that of a shallower body of the same density.

A structural feature (geological anomaly) showing relative density-deficiency, as a salt dome, gives an inverted bell-shaped anomaly normally referred to as (negative anomaly). In contrast to this case is a geological anomaly having density surplus. In this case, a positive gravity anomaly is created. The principle is shown in (Fig. 5.8).

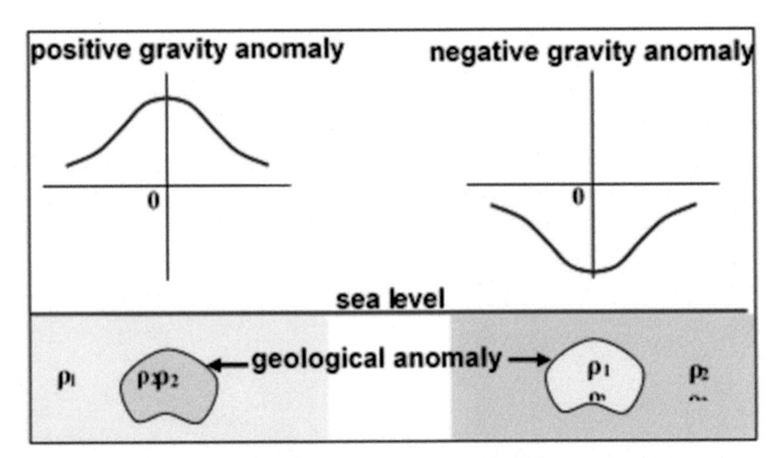

Figure 5.8 Dependence of the gravity-anomaly sign, on density contrast of two geological anomalies of densities (ρ_2 & ρ_1) in host media of densities (ρ_1 & ρ_2), where ($\rho_2 > \rho_1$).

The other factor, upon which the gravity value depends, is depth. Thus, a rock formation of a given density will give a gravity anomaly of value that varies with the depth of that formation below the observation point. A typical example for this case is the gravity changes normally exhibited over an anticline, where the gravity change (called the gravity anomaly) is greater over the anticline axis, falling to lower values over the flanks.

Acquisition of Gravity Data

The measuring instrument normally used in gravity surveying, called the gravity meter (or gravimeter), is designed to measure gravity variations rather than absolute gravity values. The gravity measurement unit (the gal) is defined to be equal to 1 cm/sec^2. Gravimeters are capable of measuring gravity changes to less than a tenth of a milligal. This means that the gravimeter is capable of measuring gravity variations to within about ten-millionth of the Earth's total gravity field which is known to be about 980 gal.

Gravity surveying is ideally suited to map subsurface rock layers or mineral deposits that show density or depth variations such as folded strata, salt domes, relatively high (or relatively low) mineral deposits, and subsurface cavities. Because of the two factors affecting the measured gravity change (density and depth), interpretation of the gravity data suffers from ambiguity in the determination of the real geological anomaly responsible of creating the measured gravity anomaly. For this reason an additional tool or information is needed to determine the true causal geological anomaly.

Processing of Gravity Data

The conventional land gravity survey is conducted by measuring the gravity value at a grid of points distributed over the survey area. The observed gravity values are then subjected to a set of corrections. The final results are gravity values reduced to values measured with respect to a fixed datum level (normally fixed at the sea level). These values are contoured to show the variation of the gravity values (in acceleration units) throughout the survey area. This contoured map (called Bouguer anomaly map) is then subjected to interpretation analysis. An example of an actual gravity survey is shown in the following figure (Fig. 5.9).

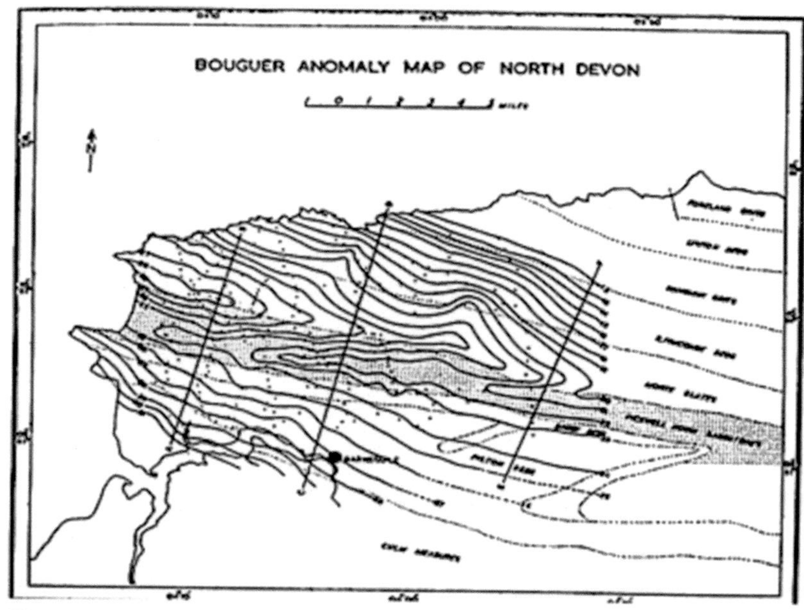

Figure 5.9 Bouguer anomaly map of North Devon (Alsadi, 1967).

Interpretation of Gravity Data

Interpretation of gravity anomaly maps is mainly based on using of the modeling approach. However, gravity anomalies may give useful indications on the subsurface structures. Geological uplifts and anticlines normally give high and positive gravity anomalies, whereas geosynclines and down-faulted rock-blocks may give rise to low anomalies. Gravity positive anomaly caused by an anticline is shown in (Fig. 5.10).

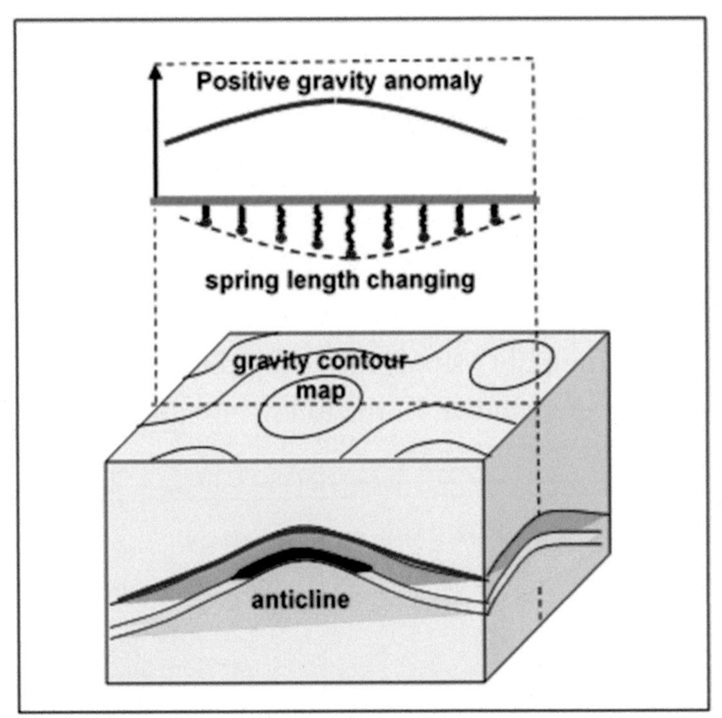

Figure 5.10 A positive gravity anomaly, caused by an anticline. Effect is schematically represented by spring-length changes.

The two types of geological features (structural and stratigraphic features) are, in principle, related to the two factors controlling the gravity anomaly; the depth and density-contrast factors. Thus, gravity values over an anticline (structural feature) are changing due to depth changes of the folded rock formations. The gravity anomaly, observed over a salt dome (stratigraphic feature), or over an igneous intrusion, is expressing the gravity effect related to the density factor. The anomaly of the low-density salt dome and of the high-density igneous plug, are negative and positive anomalies respectively.

In the recent years, aero-gravity surveying were developed and claimed to be giving satisfactory reliable results. This, of course, is much faster processing way than the conventional on-land manual surveying.

5.4.5 The Magnetic Method

A permanent bar-magnet possesses two poles, labeled as North and South poles. The similar poles repel each other while different poles attract each other. Further, a magnetic body attracts ferromagnetic materials such as iron, cobalt, and nickel. The magnetic effect can be sensed at any point in the space around any magnetic body. This is the magnetic field. A unit positive pole placed in its magnetic field will move towards the South pole as illustrated in (Fig. 5.11).

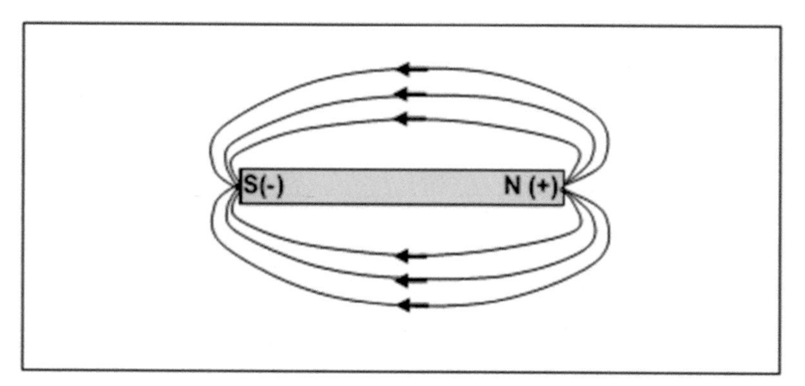

Figure 5.11 Magnetic field of a bar magnet having the two poles North (+N) and South (-S). Arrows indicate movement-directions of a unit positive pole placed in the magnetic field.

The Earth is itself a large magnetic body, having two magnetic poles. The North magnetic pole is located near the geographic North Pole and the south magnetic pole is located near the geographic South Pole. The intensity of the Earth magnetic field varies with location and with time for the same location. The field intensity also varies according to the concentration of the ferromagnetic materials in the rock cover of the Earth Crust. In general, sedimentary rocks are of weak magnetic effects on the global magnetic field. The igneous and metamorphic rocks, on the other hand, have stronger magnetic effects due to their greater contents of ferromagnetic materials.

The magnetic susceptibility

Magnetic surveying, basically depends on the variation in the body-ability of being magnetized when exposed to a magnetic field. This property (called the magnetic susceptibility), differs with different materials. Sedimentary rocks are generally of small susceptibility values compared with igneous or metamorphic rocks. For this reason, magnetic surveying is normally carried out to explore magnetized materials, such as iron-ore deposits, igneous intrusions, and surfaces of crystallized basement rocks.

Rocks which have high-susceptibility minerals, acquire magnetic intensity by magnetic induction process, creating a magnetic field which is added to the already existing Earth ambient magnetic field. As a result of this phenomenon a magnetic anomaly of a value proportional to the intensity of magnetization (of the causing geological anomaly) is created. After certain processing measures these anomalies are interpreted to reveal the causal geological structure. In modern magnetic surveys, maps of the magnetic variations are presented color-coded as shown in (Fig. 5.12).

Figure 5.12 A magnetic anomaly map is shown by color-coded anomaly levels.

Magnetic Intensity Measurements

The unit used in measurement of magnetic-field strength (magnetic intensity) is the Oersted, where one Oersted is equal to one dyne per unit magnetic pole. In practice, another smaller unit (gamma = 10^{-5} Oersted) is used. More recently an SI unit called nanotesla, where 1 nanotesla is equal to 1 gamma. The SI (Systeme International) uses MKS measurement units. The total magnetic field of the Earth is about half an Oersted.

Magnetic Surveying

In petroleum exploration, aeromagnetic surveying are usually conducted to delineate major structural changes of areas such as sedimentary basins, and regional geological changes including mapping of sedimentary basins and major rift zones. On smaller scales, magnetic surveying can be used to detect magnetic minerals such as magnetite and other iron ore deposits. The method is used to explore near-surface geological changes (such as dykes) and buried archaeological objects.

As it is the case with all of the other exploration potential methods, the magnetic method suffers from the ambiguity problem in the interpretation process. The degree of uncertainty due to this problem is reduced by using additional independent geological information.

The magnetic exploration method, like the gravity method, belongs to a class of geophysical exploration methods called (Potential Methods) because they employ natural potential fields.

Chapter-6

6. DRILLING AND LOGGING OF OIL WELLS

Drilling of the oil exploration well is considered to be the last phase of any oil exploration project. This type of exploration work provides the direct information as regards geological and geophysical properties of the subsurface rock medium. The obtained drilling data give direct recognition of the rock formations and their mineral and fluid contents (hydrocarbons and water). Associated with the drilling process, are other supporting activities which help in getting these useful data. Most important of these are laboratory analyses of the rock cuttings and well cores. Also, drilling parameters and the well geophysical logging give valuable geological and geophysical information.

The common way followed in drilling is what is known the rotary drilling which is done by use of drilling rigs, especially designed to carry out this operation. A photo of typical oil-well rig is shown in (Fig. 6.1).

Figure 6.1 An oil-well rig, in actual rotary drilling operation.

6.1 Drilling of the Exploration-Well

Since the first 69-foot oil-well (drilled by Drake in Pennsylvania, USA in 1859), drilling technology has witnessed great advances in the capability of drilling deep wells and in the control on the direction of the drilled well. Inclined, and even horizontal, drilling is now in common application.

Drilling of the first oil well after completing the geophysical surveying is called (Exploration drilling) or (Wildcat drilling). There are other types of drilling labeled according to the purpose of the drilled well. For instance, it is called Production drilling when the well is designed for normal oil production, and Observation drilling for monitoring the behavior of the oil reservoir. Water- (or gas-) injection drilling is special for injecting water (or gas) to boost the reservoir pressure and enhance oil recovery. An exploration well provides important information, thus:

- Drilling parameters (like drilling rate) give indications on the physical nature of rocks penetrated by the drilling process.
- Rock cuttings and cores provide direct information on the rock lithology and thicknesses of the penetrated rock formations.
- Cores, and cuttings, and the circulating drilling mud, give direct knowledge on presence of hydrocarbons.
- Analyses of the well logs (as electrical logs, radioactivity logs, and sonic logs) furnish valuable information on the subsurface geological nature and on the hydrocarbon contents.

6.1.1 The Oil-Well Rotary Drilling

The common technique applied in drilling an oil well, is what is called (Rotary Drilling). The drilling process is done by rotating a column of steel pipes, with a drilling head (called the Bit) attached to its lower end. By its rotation, the bit carves a hole through the rocks, and, in this way, a drill-hole is made. Drilling mud is pumped through the pipe, returning to surface via the space between the drill pipe and well wall. Mud circulation will carry rock cuttings to surface and provide cooling and lubrication to the drilling process. Rotary drilling of an oil well is summarized as follows:

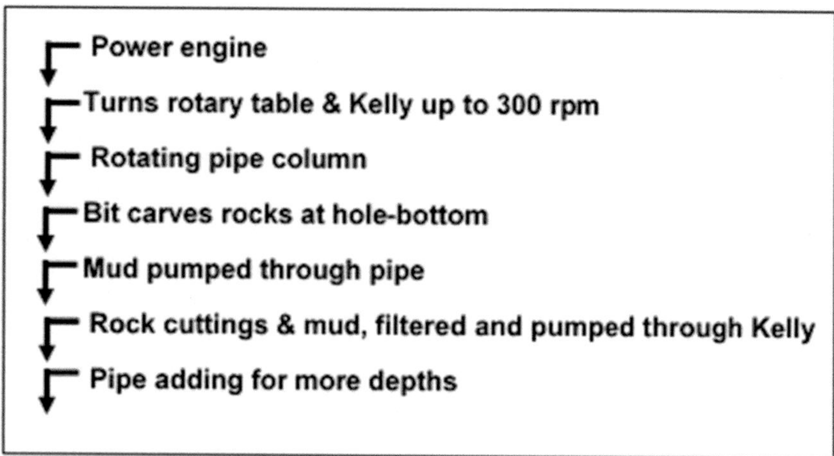

Summary of the operation sequence followed in rotary drilling

Drilling operation is done by use of a mechanical steel structure designed in the form of a tower of height of (120 feet – 150 feet) and of strength that can handle a load of up to 300 tons (Wymer, 1964, p.46). This huge drilling system is called drilling Rig. A schematically representation of a typical oil-well drilling rig, with its principal parts, is shown in (Fig. 6.2).

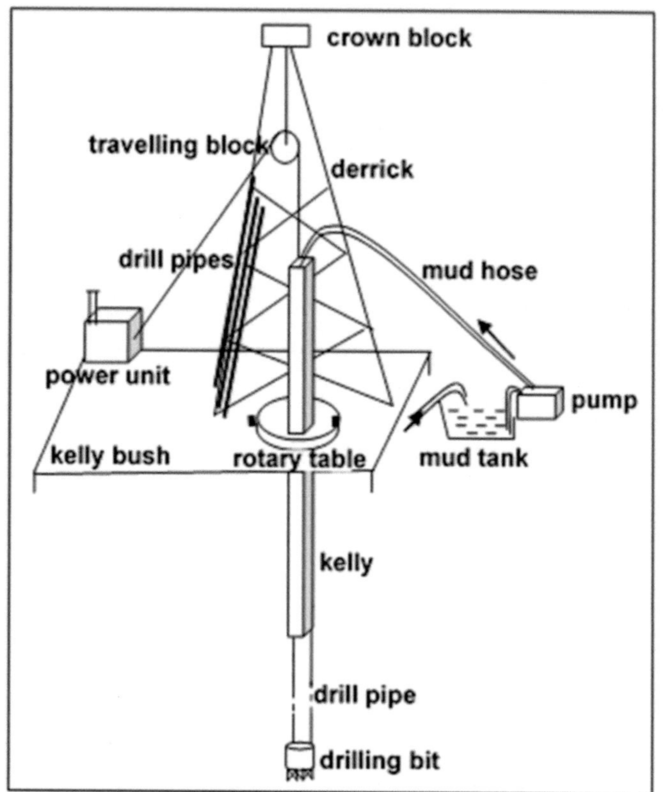

Figure 6.2 A Simplified sketch showing the main parts of a rotary drilling rig.

The drilling rig, shown in (Fig. 6.2) above, is made up of the drill tower (called derrick), drill pipes, drilling floor (Kelly Bush), and power unit to provide energy for rotating the pipe column and to circulate the drilling mud. The grinding process is executed by the rotating bit attached to the end of the drill pipe.

The more common types of bits are: (grinding bits) which produce rock cuttings that are removed by the circulating mud, and (coring tubular bits) for producing rock-cores.

6.1.2 The Drilling Mud

During rotation of the bit, a specially provided pumping unit pups the drilling mud fluid which consists of suspension of special mud material mixed with water (or with oil as it is sometimes done). The mud is pumped through the drill pipes, the bit, and then returns to surface through the space between the drill pipes and the wall of the well. On arrival at the well head, the mud fluid, with the rock cuttings it contains, is collected and the cuttings are separated for examinations later on.

Drilling mud is essential in the drilling process as it provides cooling and lubrication for the rotating bit during operation. Due to the mud column pressure which is acting on the well-walls, pressure of fluids (gas, oil, and water in the penetrated formations) is counteracted and hence preventing into-well infiltration of these fluids. Also, mud is preventing any possible blow-out which may happen if an abnormal high pressure is met-with. Control of mud pressure is done by mud-density manipulations.

Another important benefit of the drilling mud is prevention of caving by the formation of a thin mud layer (called the mud cake) which is smeared over the well walls. Mud circulation helps smooth drilling operation, cleaning the well from the debris (rock cuttings) resulting from the abrasive action of the rotating bit. Also, the lubrication effect of the mud helps avoiding of possible stuck-pipe events and other such like problems.

Geologists collect, document, and study the rock cuttings, brought to surface by the circulating mud.

6.1.3 The Drill-Hole and Well Casing

During the drilling process, drilling mud is circulating carrying with it the produced rock cuttings. According to the extent of penetration into the well wall of drill-fluid, three types of zones are recognized (Fig. 6.3):

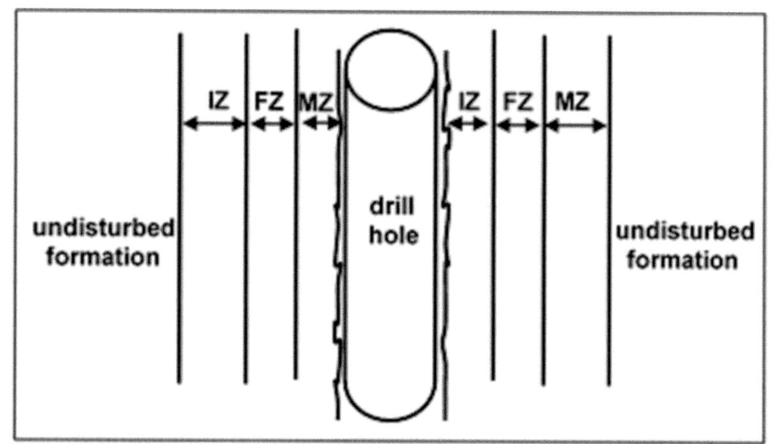

Figure 6.3 Mud invasion-zones in the well wall, labeled: mud-cake zone (MZ), flushed zone (FZ), and invasion zone (IZ).

Mud-Cake Zone (MZ)

This is a thin coating of well-wall which is of few tenths of an inch.

Flushed Zone (FZ)

Natural fluid is completely replaced by the drilling fluid (several inches).

Invasion Zone (IZ)

Drilling fluid infiltrated into region surrounding the well (few inches as in shale to several feet as in porous sandstone).

After removing drill-pipes, a pipe column is lowered into well, then cement is pumped-in to hold it tight in place. Casing prevents caving and fluid seeping. However, to allow oil and water to flow into well, the casing is later-on perforated at the appropriate places.

Electric logs cannot usually be run with casing and sonic logging is severely disturbed. In general, nearly all logging processes are carried out in uncased (open-hole) wells.

6.1.4 Role of the Geologist in Well Drilling

An oil well drilling project is basically a team-work operation in which geologists and engineers work together in completing the drilling project. In general, engineers take over the task of execution of the drilling process, while geologists participate in decisions as regards drilling parameters and in interpretation of the data obtained from the well. The geologist role in the drilling project can be divided into the following three stages:

Stage-1, Planning and Design of the Drilling Program

According to the available geological and geophysical data, the geologist determines the well location and its total depth. He participates with the engineers, in the design of the drilling program which includes depths and specifications of the well casing. The design program includes also work plans for obtaining the rock cores and cuttings in addition to defining of the program for the well geophysical logging.

Stage-2, On-Site Geological Control

This phase of work starts as soon as the drilling operation begins. The on-site geologist has to actually live by the well throughout the drilling period. He shall keep a direct contact with drilling work, observing and following up the drilling progress and documenting all of the drilling progress and reporting his observations regularly to his superiors. In particular, he needs to report the depths and thicknesses of the penetrated formations, oil shows, and any drilling problem, a drilling operation may meet.

The geologist responsibility is to collect and clearly document the rock cuttings after being washed and dried. The collected cuttings are sent to the central laboratory for analyses and palaeontological and palynological studies. At the well site, chromatography analysis is usually conducted to test for presence of hydrocarbons.

Stage-3, Monitoring of Coring and Core Samples

By use of special types of drilling bits, the geologist takes the decision as regards the parts of the well for which core samples are to be taken. This decision is usually taken in collaboration with the reservoir and drilling engineers. The site geologist usually supervises all activities concerning rock cuttings and core samples and closely watches the drilling operation including mud-fluid circulation and the well logging at a later stage.

The core samples give valuable information on the macrofossils in addition to the microfossils contents. Also these samples can be subjected to petrophysical analyses to determine physical properties (as porosity and permeability) and formation dips. Compared with the cuttings, cores give more accurate depth and boundaries of the penetrated formations.

6.2 Well Geophysical Logging

After completion of the drilling operations of a well, a group of technical activities (well logging) are done to extract direct information on the rock formations penetrated by the drilling process. Study of the penetrated rock formations (commonly referred to as formation evaluation) includes examination of the circulated mud (mud logging) and study of its contents of rock cuttings, detection of hydrocarbon matter, and documentation of lithological changes, and fossils contents. The greater bulk of activities done at this stage is the geophysical logging of the drilled well which is always done concluding the drilling operation of any drilled exploration well. Well logging give direct determination of petrophysical properties of the penetrated formations, which are essential in the formation evaluation and reservoir characterization.

The well log (or wireline log) is recorded by a special well logging tool (called sonde), carrying sensors which are lowered into the hole by a cable. The standard procedure is to start the measurements at the bottom of the hole and moved upwards through the borehole. Measurement of a certain geophysical parameter is done either continuously or at discrete points. The output is recorded and normally produced as charts showing the measured value as function of well-depth, as it is sketched in (Fig. 6.4).

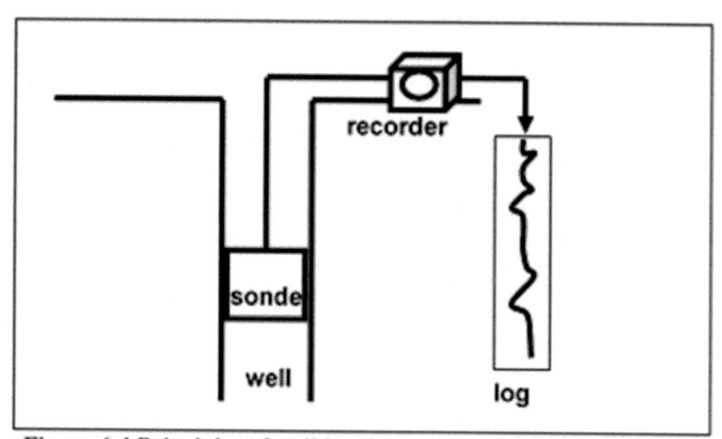

Figure 6.4 Principles of well logging

The well logging process results in measurements-data normally plotted as charts called (well logs). Since it was introduced (by C & M. Schlumberger in 1928) the geophysical well logging technique has vastly developed into an indispensable formation evaluation-tool for the rocks penetrated by drill-holes. Logging methods can be divided into the three principal methods; electrical-, radioactivity-, and acoustic-logging.

6.2.1 Electrical Logging

Basically, electrical logging involves measurements of the variations of electrical resistivity and natural potential of rocks down the drilled well. Depending on the applied electrode configuration, the following techniques are in common use.

(i) Electrical Resistivity Logging

This type of electrical logging is based on a configuration whereby a DC (or low frequency AC) source electrode is lowered down the drill-hole and measuring the potential drop across a set of potential electrode. The output is a continuous record of the variation of the electrical potential (or the corresponding apparent resistivity), with depth. This method must be applied in uncased wells.

Ohm's Law (V=IR) which states that the electric current (I) is proportional to the electric potential (V), with the resistance as the proportionality constant. Thus, from current and voltage measurements, resistance can be computed. For well resistivity logging, there are two types of configurations, the normal-sonde and the lateral-sonde logging (Fig. 6.5).

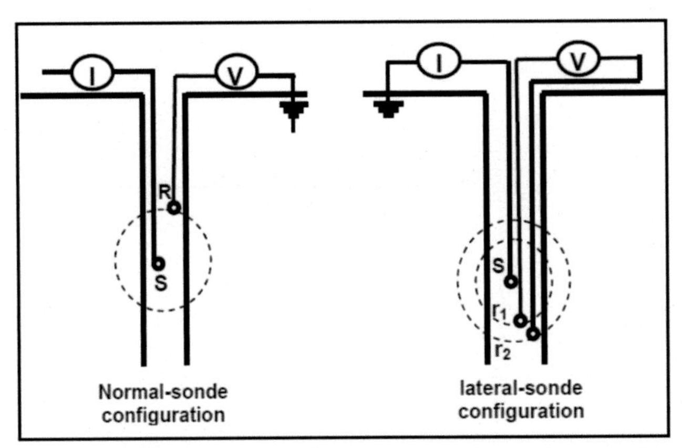

Figure 6.5 The two types of electrode configurations used in resistivity well logging; normal sonde and lateral sonde configurations.

The resistivity log expresses measurements of the electrical resistivity in the usual (ohm units) for the rock medium surrounding the drill-hole. Penetration distance of the electric field in the penetrated rocks depends on the electrode spacing-distance. The larger the electrode spacing, the greater is the penetration distance (Fig. 6.6).

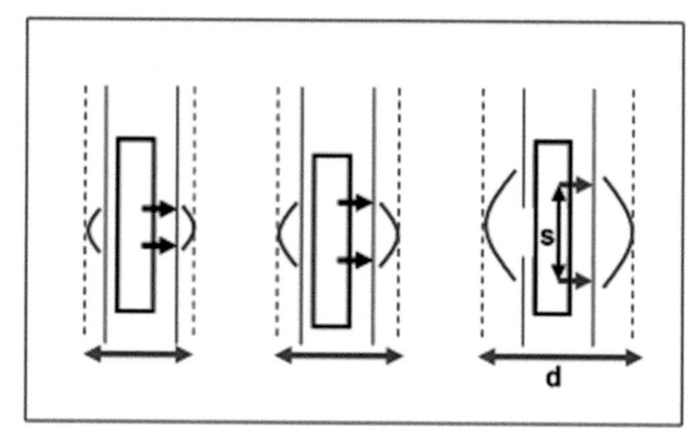

Figure 6.6 Lateral distance (d) penetrated in resistivity logging is proportional to electrode spacing (s).

The normal sonde electrode-configuration includes two electrodes; one source and one receiver, spaced few feet apart. The instrument reading which is given in apparent resistivity units reflects the properties of the region near the source-electrode. The other type (the lateral-sonde configuration) is equipped with three electrodes; one source and two receivers. A variation was made on the lateral sonde whereby the current rays spread out horizontally (focused rays) rather than radially (unfocussed) spreading. This modified procedure is called (laterologging) and the produced resistivity chart is called (laterolog). The lateral logs give more accurate and sharper boundary detection, in addition to reduction of effects of the mud and hole-diameter variations.

Resistivity logs help in the diagnosis of types and boundaries of formations. For example, low resistivity indicates higher porosity and permeability of water saturated formations, whereas increase of resistivity can reflect existence of oil and gas. In general, low-porosity rocks (as shale) and porous rocks saturated with salty water exhibit low resistivity. On the other hand, porous rocks saturated with low-salt fresh water, or saturated with oil will exhibit high resistivity. On this basis, resistivity logs serve as important and effective indicators for presence of hydrocarbons.

(ii) Electrical Induction Logging

The logging sonde uses coils instead of electrodes. A primary coil carrying an AC current creates an alternating magnetic field which is inducing currents into the rock formations. These currents, in turn, create secondary magnetic field which induces, in the receiving secondary coil, an AC which varies with the resistivity of the formations. The coil-sondes configuration is shown in (Fig. 6.7).

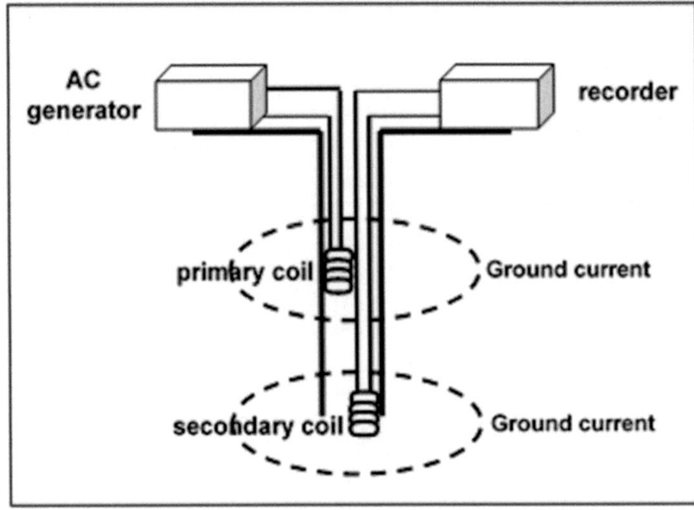

Figure 6.7 Coil-sondes configuration used in induction logging

(iii) Spontaneous Potential (SP) Logging

This logging method depends on measurements of the natural electric potential (usually in millivolt units) of the rock medium surrounding the surveyed well. It is normally referred to as self-potential or spontaneous potential (SP) logging. It is used only in an uncased holes filled with conductive mud. The measuring sonde is of simple configuration. It consists of only two electrodes; one is lowered into the well by an insulated cable and the other is fixed at the ground surface (Fig. 6.8).

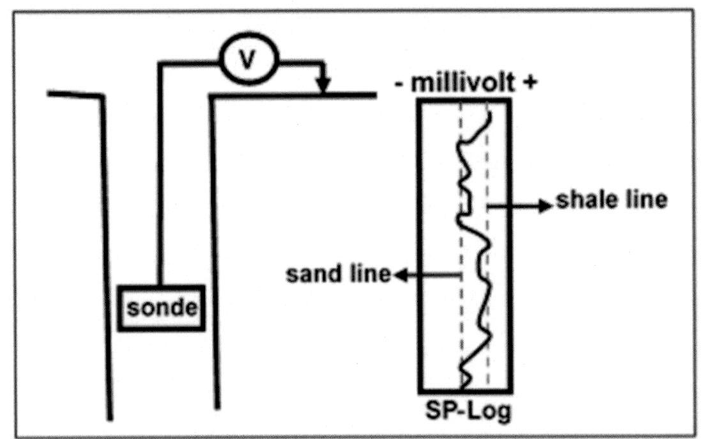

Figure 6.8 Configuration used in spontaneous potential logging with schematic representation of the SP log

Interpretation of the SP logs depends on the manner of the electrical potential variation. As a general rule, changing of the log values towards positive potentials is considered as an indicative of impermeable rocks as shale or tight limestone or tight sandstones. When the change is towards negative potential, it is interpreted as being due to porous sandstones or porous limestone. In this way, the SP logs give useful indications on lithology, water salinity and help in determination of formation boundaries. In particular, it shows the boundaries of formations in sand-shale sequences, which are determined with the help of drawing lines (sand and shale base-lines) in the produced SP log as shown in (Fig. 6.8) above. It is commonly observed that shale beds give the same level of SP-readings allowing for drawing a straight line indicating shale SP-value. This is called (the shale line). Similarly a (sand line) can be drawn for sandstone SP-value. The inflection points in the SP-log indicate formation boundaries.

This method, which needs no artificial current, can detect natural potential differences which develop at formation boundaries. Compared with resistivity logs, the SP logs give more sharp changes and hence more accurate formation-boundaries. Like lateral logging mentioned above, there is a variation made on induction logging to give focused current radiation for getting sharper boundary indication. Induction logging is used in wells filled with no-conducting drilling-mud. It can as well be used in mud that is conducting.

6.2.2 Radioactivity Logging

Radioactivity is the phenomenon of emission of particles and photons of electromagnetic energy from an atom. This radiation process occurs either naturally from unstable nuclii or induced by bombarding of stable nuclii by photons or particles.

Examples of such radioactive elements are uranium, thorium rubidium and potassium 40 which is most commonly found in shale and clay and less in sandstone and limestone.

There are three types of radiation; Alfa particles formed of charged helium nuclii, Beta particles formed of high speed electrons, and Gamma rays of electromagnetic wave-energy. Out of these, only gamma ray is used in well radioactivity logging. The detection instrument is a type of a scintillation counter which consists of a special crystal (like sodium iodide), which emits flashes of light as they absorb gamma-ray photons, hence the name (scintillation counter). A photoelectric tube converts these flashes into electric currents which are displayed in the form of a continuous chart (the radioactivity log).

Radio activity logs provide important information on rock lithological types, especially on those containing certain concentrations of radioactive minerals. Thus, these methods are ideal indicators of shales and clays which, by nature, contain radioactive minerals in their makeup.

Three methods of gamma radioactivity logging are in common use. These are: natural Gamma-ray logging, Gamma-ray rock-density logging, and neutron Gamma-ray Logging methods.

(i) Natural Gamma-Ray Logging

This method uses a detector mounted on a sonde to measure the naturally emitted gamma rays from the radioactive minerals existing in the rock formations. Unlike electrical logging, gamma-ray logging can be run in cased wells, as well as in uncased wells, with detection penetration of few feet from the well walls.

The resolution power of formations boundaries is affected by the counting time of the instrument and sonde logging speed. Reasonable results are obtained with a counting time of 2 seconds and sonde speed of 150 mm per second. Measurements can be made in cased wells, but the intensity of radiation is reduced by about 30% in this case (Kearey and Brook, 2002, p.244).

In general, high log values are interpreted as increase of the shale percentage, while low values are considered as indications for sandstones and limestone rocks. To aid interpretation, it is possible to draw shale-lines and sand-lines on the log chart. An advantage of the gamma log is its capability of differentiating between shale and sandstones, independent of the porosity and permeability characteristics of the rocks.

(ii) Gamma-Ray Rock-Density Logging

The sonde, in this case, contains a gamma-ray source and a scintillation counter to detect the gamma ray which is back-scattered from the formations and received by a detector fixed at a certain distance from the source.

The gamma-ray photons collide with the electrons of the elements in the formation resulting in loss of photon energy and back scattering of gamma-ray which has a wavelength different from that of natural gamma-rays. This is called (Compton Scattering Effect). The collision rate and the back-scattered (secondary radiation) are proportional to electron density which is, in turn, proportional to rock-formation density.

In practice, the gamma-ray detector is shielded to record only the secondary radiation, and the sonde is firmly pressed to well-wall and moved slowly (less than 30 feet/min) in order to maintain good contact.

A variation to the method is introduced to provide corrections for mud-cake effect. A secondary detector is included in the sonde which is responding to mud-cake and small wall irregularities. The resulting log (called Compensated Density Log) shows both of the bulk density (ρ) and the density-correction log ($\Delta\rho$). The produced chart expresses the formation-density log.

The principle of this type of logging (sometimes called (gamma-gamma logging) is shown in (Fig. 6.9).

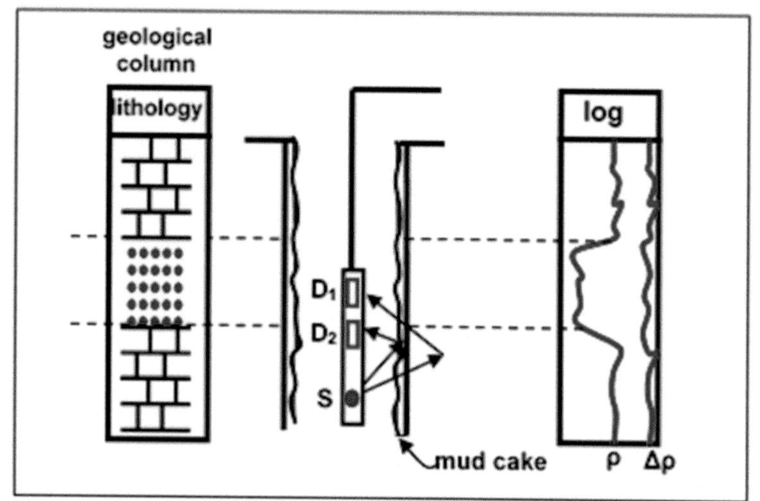

Figure 6.9 Principle of Gamma–gamma log recording. Detectors, (D_1) and (D_2) record secondary and mud-cake radiations respectively. The symbol (S) represents the source.

Interpretation of the density log is based on the direct proportionality existing between the recorded scattered gamma-ray intensity and the number of electrons found in the scattering rocks. The number of the scattered electrons is, in turn, proportional to rock bulk density.

The Gamma-gamma log can be used in computing porosity (Ø) by applying the relationship:

$$Ø = (\rho_m - \rho_l) / (\rho_m - \rho_f)$$

where, (ρ_m), (ρ_l), and (ρ_f) represent matrix density, log read-density, and pore fluid density respectively.

(iii) Neutron Gamma-Ray Logging

The sonde consists of a neutron source and detecting scintillation counter at a fixed distance apart. The source is a small radioactive body (such as plutonium-beryllium) which emits neutrons during the process of radioactive decay.

When the source-generated neutron collides with a hydrogen nucleus (which is of a matching mass) its kinetic energy is reduced to an extent that it can be captured by a large nucleus (as the hydrogen nucleus) causing emission of a secondary gamma radiation (capture gamma radiation). The principle of the logging is sketched in (Fig. 6.10).

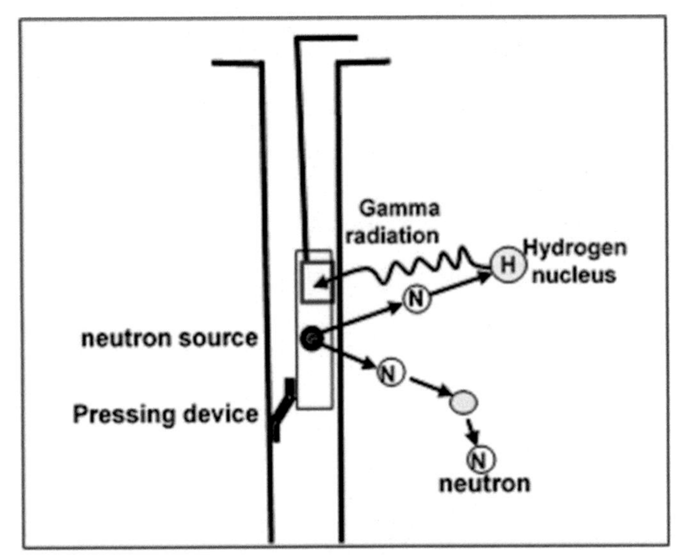

Figure 6.10 Principle of Neutron Gamma-Ray logging.

In the logging process, the sonde is moved at a speed of 30 ft/min with 2 seconds for the counting time. A skid (pressing device) is provided to keep the sonde in close contact to wall. The gamma radiation, generated by the capturing-phenomenon, comes from the material surrounding the drill-hole. The produced log (which can be run in cased or non cased well) is displayed in the appropriate measuring units or in porosity percentages.

The intensity level of measured radiation is proportional to concentration of hydrogen, which is existing in water, in hydrocarbon, and in hydrous minerals such as silicate-clays, micas, amphiboles, and gypsum. Thus in carbonates hydrogen source is water, and in sandstones, hydrogen source is in hydrocarbon found in the pores of the rock. In shale however, mica and clay minerals contribute to hydrogen content as well as from pore water. In this case other types of logs (e.g. gamma-ray logs) are needed to distinguish shale from water-saturated porous sandstones or from water found in fractured limestone.

In general, best application of neutron logs is in following up the porosity variation of porous rocks using the direct proportionality between porosity and gamma-ray intensity-level. In short, this log, which depends on the gamma ray generated from neutron bombardments of hydrogen atoms. The generated gamma ray is proportional to concentration of the hydrogen element in the penetrated rocks, and, in turn, on the water and hydrocarbon fluids in those rocks. This also means that high log values indicate high porosity.

It is useful to note that in case of presence of hydrocarbon gas in high-porosity formations, density logs (gamma ray rock-density logs) are expected to give low log values compared with the neutron log vales at the same zone, where the neutron readings expected to be relatively high. This means that both of the logs (density and neutron logs) are needed in order to detect hydrocarbon gas.

6.2.3 Acoustic Logging

The purpose of this type of logging is basically for getting information on velocity of propagating acoustic (seismic) waves. Sonic logging, well velocity surveying (well shooting) and VSP surveying are included in this type of well-logging.

(i) Sonic Logging

The logging sonde, in its standard form, consists of two receivers about one foot apart and a source at about three feet from the nearest receiver. To correct for tilting and hole-irregularities effects, a dual source sonde is used, making what is called a borehole-compensated sonde (Fig. 6.11).

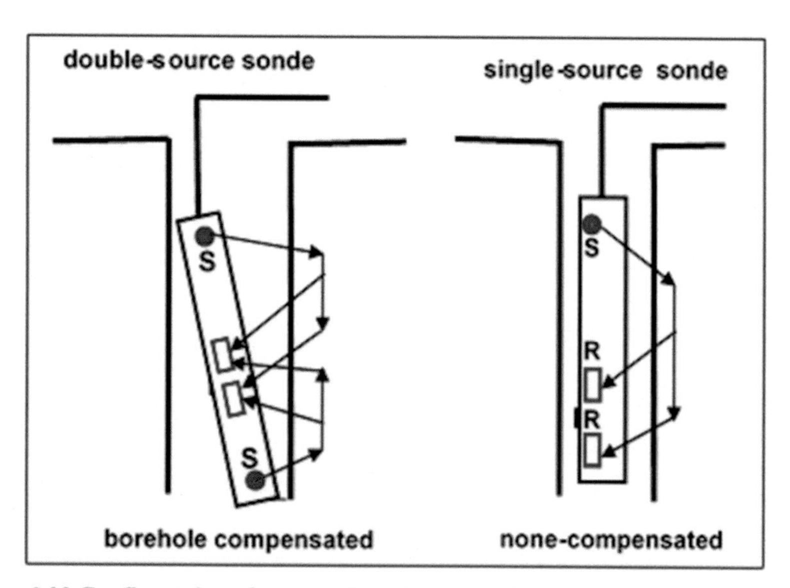

6.11 Configuration of source (S) and receiver (R) of the sonde employed in sonic logging.

The electronic structure of the sonde is designed in such a way that the output is made to be the difference in the travel-times to the two receivers. The time difference, measured in time-units per one foot, called (interval transit time), is plotted (normally in micro seconds) against depth to give the continuous wiggly curve known as the (sonic log). In the compensated sonic logging, seismic pulses are emitted alternately from the two sources and the transit times from the two oppositely traveling refracted P-waves are averaged electronically. The output (transit time) is plotted against depth, giving the borehole-compensated (BHC) sonic log (Fig. 6.12).

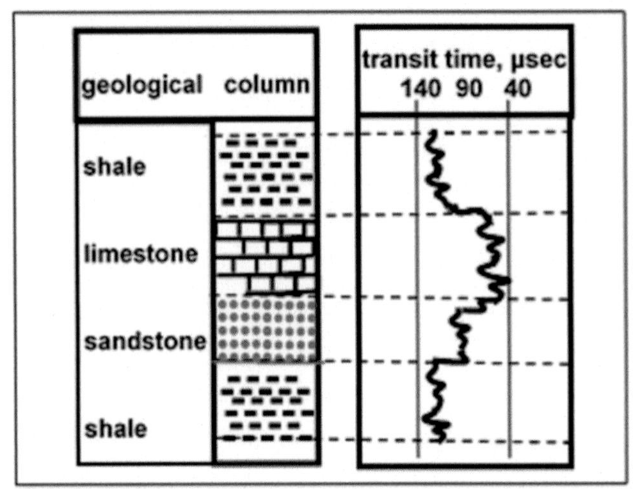

Figure 6.12 Sketch of sonic log for a hypothetical geological column.

The borehole compensated sonde (BHC) gives an average interval transit time which is plotted on a paper strip. The produced log in this case is normally referred to as BHC sonic log which is used to identify lithologies, determine formation boundaries, and in computing synthetic seismograms. The interval transit time can be integrated down the well to give the total travel time. This type of logging can only be run in an open (uncased) hole.

The interval transit time can be used in computing porosity (Ø) by applying the relationship:

$$Ø = (\Delta t_l - \Delta t_m) / (\Delta t_f - \Delta t_m)$$

where, (Δt_m), (Δt_f), and (Δt_l) represent transit times of matrix material, pore fluid, and log read-transit time respectively.

(ii) Well Velocity Surveying

A hydrophone-type detector is lowered down the well which is filled with the drilling fluid. The travel time of the seismic wave generated by a surface-placed shot and received by a detector placed at a formation boundary is recorded. The full log is obtained by repeating the recording at each boundary traversed by the well. From the measured travel time of direct arrivals, the average velocity and interval velocity are plotted as function of well-depth. The velocity survey is also called (check-shot survey). Principle of the velocity survey setup and velocity-depth plot is shown in (Fig. 6.13).

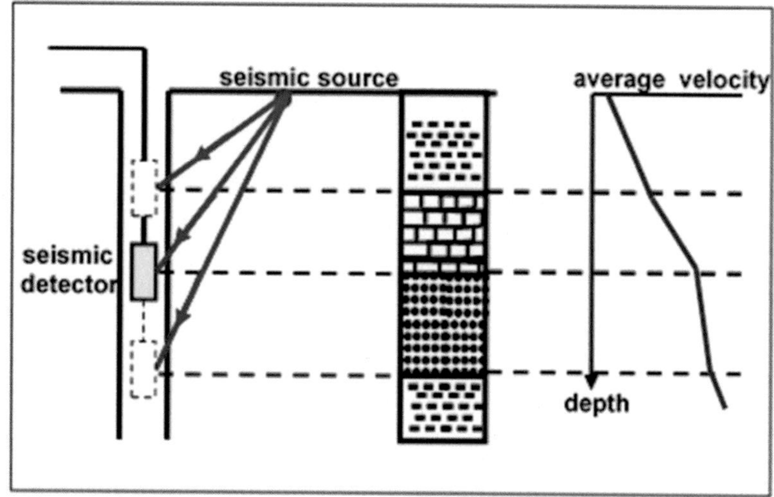

Figure 6.13 Principle of the well velocity surveying and a typical velocity-depth plot obtained from the survey.

The obtained velocity information is used to calibrate the sonic log and check the sonic-log integrated time (hence the name, check-shot surveying).

(iii) Vertical Seismic Profiling (VSP)

Basically the measurement set-up is the same as the check-shot recording system. The difference is in the recoding duration time which is here extended to allow recording reflected waves as well as the direct arrivals. At each detector stop-location (normally at 25 m spacing), the recorded seismic trace is allowed to include events from up going reflected waves in addition to the first arrival (direct wave) event. The terms (down-going wave) and (up-going wave) are normally applied to refer to the direct and reflected waves respectively. The ray paths of the VSP shooting and the corresponding travel-time plot of the recorded waves are schematically shown in (6.14).

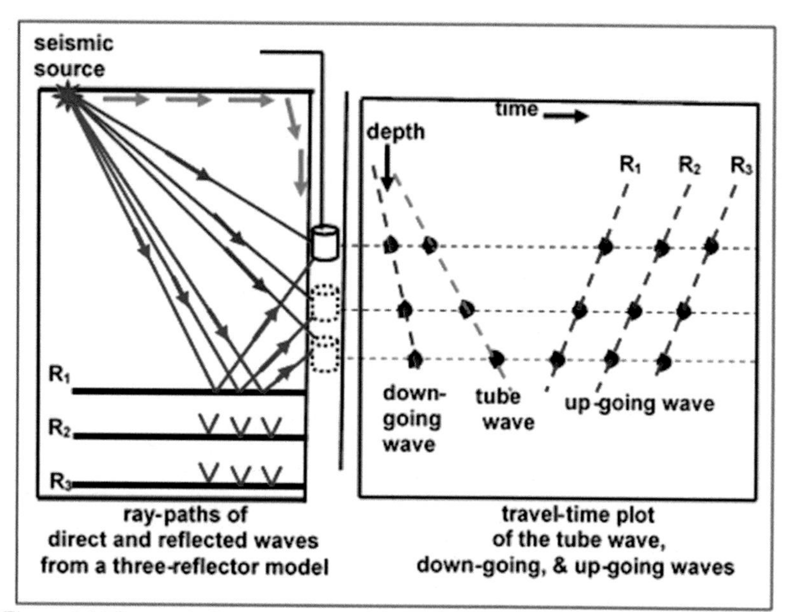

Figure 6.14 Ray-path diagram and the corresponding travel time plot of a VSP survey of a geological model made up of three reflectors.

A seismic pulse is generated on the surface from dynamite explosion or from an air gun submerged in a water filled hole. This is repeated at all of the hydrophone positions, normally positioned at 25 m-spacing down the well.

With an appropriate processing sequence, the final corrected VSP section is produced. Processing includes, data editing, correction to vertical time and velocity filtering for separating unwanted events. It is evident from the VSP section that downward events (primary and multiples) increase in time with depth, and the upward events (primary and multiples) decrease in time with increasing depth. By arranging the produced seismic traces (25 m spaced down the hole) a seismic VSP section is obtained. Each seismic trace of a VSP section contains events from down-going waves (direct and multiples arrival) and from up-going waves (reflections and multiples). The main application of VSP is providing seismic section of reflectors which have not been reached by the drilling

(iv) Other Logging Techniques

Some logs are used to give geometrical information on the deviation from the vertical (hole-drift angle) as well as the azimuth of the deviation. Other logs are made to give the bedding dip (dip-meter logs) and measurements of the well-hole diameter (caliper logs). On more limited scale of application, other types of well logging have been used. Specially modified borehole-gravimeters, magnetometers, and thermometers are examples of such tools. Geothermal prospecting is applied to detect geological features which affect heat flow such as shallow salt domes, faults and dykes. Also ground water is investigated by this method.

6.2.4 Log Interpretation

The process of analysis and interpretation of well logs is sometimes called (formation evaluation). Normally all the logs obtained for a well and their analysis-results are presented in one combined display called the (composite log). This comprehensive log usually contains all (or most of) the following log data:

- Geological column showing lithological and palaeontological information
- Borehole compensated sonic log
- Gamma-ray and Neutron logs
- Resistivity and SP logs
- Drilling information (drilling rate, mud density variation)
- Caliper and dip-meter logs.

The main information which can be obtained from well logs can be summarized as in the following table:

Log Type	Application	Comment
Electrical Logs - Resistivity - Induction - Spontaneous Potential	- Fluid-type identification - Porosity evaluation - Boundary determination - Shale and sand lines	Uncased well
Radioactivity Logs - Natural gamma-ray - Gamma-ray Density - Neutron gamma-ray	- Shale/sandstone - Formation density - Porosity - Fluid type	Cased well
Acoustic Logs - Sonic logs - Well velocity - VSP	- Lithology type - Boundaries - Synthetic seismic -Seismic-velocity -Reflection identification	Cased & uncased wells

6.3 Well Completion

According to the planned drilling program the well drilling operation is done in several stages. At the first stage, and to prevent caving-in of the well walls, drilling is stopped at a certain depth and a steel pipe (called the casing pipe) is lowered into the well and fixed in position by pumping cement mixture in the space between the casing pipes and the well walls (cementing process). In the next stage, drilling is continued with a smaller bit-size to the planned depth. At this stage, the drilled hole is lined by the appropriate casing which is likewise cemented in. The drill-casing-and-cementing phases are repeated until the final planned total depth is reached. The end result of the casing operations is a set of concentric pipes each of which is ending at the surface.

When the planned total depth is reached, all used drill pipes including the attached drilling bit are withdrawn. The next operation is perforation of the parts of the casing facing the oil bearing formations. The so-produced holes are made to allow the oil to seep inside the central production tubing of the well.

The last stage in preparing the well for the production process is fixing of a group of flanges which provide a sealing structure connecting together all the surface ends of the casing pipes. This part of the well which contains pressure equipment in addition to the mechanical system that provide the blowout-preventer (BOP) is called the well head which is also equipped with an assembly of monitoring gauges and valves that controls flow, and any other possible interventions. This assembly is normally referred to as the Christmas tree (Fig. 6.15).

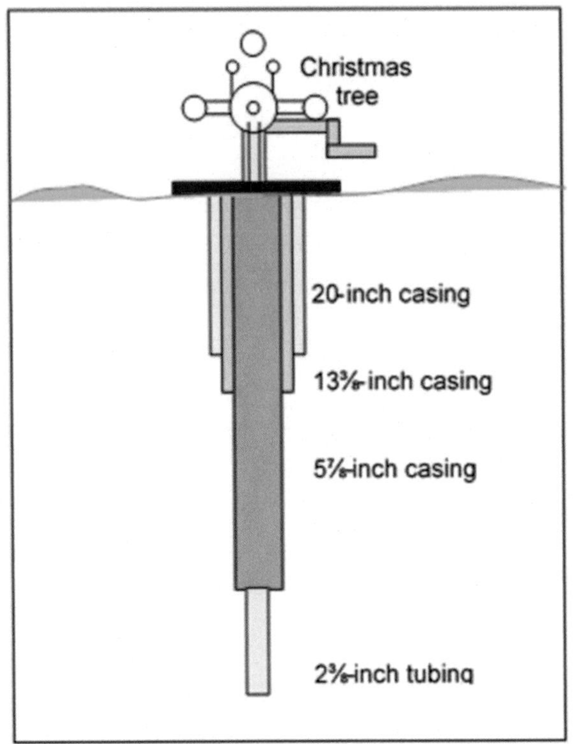

Figure 6.15 Typical casing statistics of a drilled well after casing and installment of the well head monitoring gauges (the Christmas tree).

To summarize; well completion is the process of making a drilled-well ready for oil production, or for fluid (water or gas) injection. The process principally involves, running in the central production tubing and perforating and stimulating to allow oil to seep in into the production tube. Sometimes, the cementing process of the casing is considered as integral part of the completion process.

Chapter-7

7. FIELD DEVELOPMENT AND OIL PRODUCTION

Oil production is that part of the oil industry project which concerns extraction of the crude oil from the oil reservoir. It covers all other activities for carrying out gas and water separation processes which secure oil in a state ready for transportation to the oil refinery or to the export terminals.

After the exploration activities and drilling of the first exploration well are over, preparations for the following stage of the industry is started. At this stage oil-production facilities are designed and installed with the specifications which comply with requirements of the newly discovered oil field. This process includes well casing, well completion, and well-head installment. Data from the exploration and appraisal wells form the basic requirements for the needed field development activities.

7.1 The Oil-Field Development Plan

At this stage, a detailed development plan for exploiting the discovered oil field is established. This plan is based on the following specifications:

- Properties of the crude oil, its gas-oil ratio (GOR), and sulfur content
- Reservoir pressure and closed-in wellhead pressure
- Relationship between oil flow-rate and pressure decrease
- Oil zone parameters, as pay-zone oil column, gas-oil contact (GOC), and oil-water contact (OWC).

These data are used as basic guides for drawing the field development plan, and to enhance production efficiency, in addition to design of the appropriate production equipments and drilling program of the required production wells.

7.2 Oil Field Development-Activities

The conventional approach in field developments of onshore oil fields starts with locating the well sites in which to drill the required production wells with the pipeline network conecting these wells. With time, and according to accumulation of future field-data, modifications and more technical activities may be done. Detailing exploration activities, drilling additional appraisal and evaluation wells, may be carried out for the purpose of production or for water (or gas) injection if, and when, needed.

In the case of offshore oilfields which happen to be located in deep-water (100m-200m), a production platform is constructed to house the production equipments, operation personnel, and their living facilities. The development activities include provisions for transporting the produced oil to a tank-farm established near-shore land site, if no such storage facilities were provided on the near-the well platforms. In general, field development of an offshore field needs more advanced and higher cost installations compared with the onshore oil fields.

7.3 The Reservoir Energy

The main sources of energy that activate oil-production drive-mechanism come from the fluid pressure acting on the oil body. Two principal factors are forming the reservoir energy that enables the crude oil to be pushed up the well with no artificial aiding facilities. The first factor is the pressure of the formation water existing below the oil accumulation. The water drive process represents the principal agent that keeps the production continuity. The second factor contributing to the reservoir energy is the gas cap formed above the oil body as a result of liberation of the in-oil dissolved gas. The gas cap provides pressure on to the oil, pushing it into the direction of the lower-pressure parts of the reservoir located near the oil-producing well. In fact, both of these two factors (the underlying water and the overlying gas) are jointly providing the active reservoir energy that keeps the production process.

In consequence to oil depletion, the oil-water contact (OWC) is upward-shifted and the gas–oil contact (GOC) is downward shifted. Under certain conditions, the water (or the gas) meets higher permeability flow-paths through which these fluids can find their ways to the well. In such a case, the well starts to produce water or gas instead of oil. Petroleum engineers call this phenomenon (channel flowage, or channeling). This is considered to be a serious problem that needs an appropriate solution. The usual way to restore oil production is by making extra perforations, in the casing of the well, opposite the oil bearing zones.

7.4 Effect of Production on Reservoir Energy

As a common observation on the reservoir behavior, it is noted that, during the early years of production, oil is flowing out by its own internal energy. However, with continuous depletion, reservoir energy gets reduced to an extent that the energy becomes insufficient to push the oil column up to the well head. When this low-pressure situation is reached, certain energy-boosting measures need to be taken in order to restore oil flow from the wells. Among the ways followed in increasing reservoir energy and to enhance oil recovery is water injection, or gas injection.

On start of the production operations, pressure of the dissolved gas is at its maximum level, but with the on-going production, the pressure starts to decline until it (the saturation pressure) reaches a level when the gas starts to liberate from oil in the form of gas bubbles to form a gas cap over the oil body. A side effect of this process (gas separation process) the viscosity of the oil increases. Consequently, the more viscous oil becomes of less capability to move which leads to decrease in the overall production efficiency.

With continuous oil production, another important effect takes place to the reservoir energy. This is the decrease in the energy of the formation water. The water energy gradient in the up-dip direction decreases, giving way to the gravity effect which acts in opposite direction (down-dip direction). In water-drive reservoirs, the main source of the reservoir energy comes from the formation-water pressure which usually stays in effect throughout the production.

Another readily observable reservoir-change, resulting from continued production, is lowering of the oil-water contact which changes at a rate proportional to the number of the active production wells. At this point, it is worth noting that there is an optimum depletion rate (or maximum efficient rate, MER), with which the reservoir is safely produced with increase of the reservoir production age. Too-high depletion rate leads to increase of oil inter-pores motion and creation of cut-off pockets of oil bodies which may not be possible to recover.

7.5 Reservoir Behavior during Production

During production, petroleum engineers are monitoring reservoir behavior through studying fluid volumes, pressure and temperature changes. Studies of these interacting reservoir parameters is usually done through a special equation called (Material Balance Equation). The collected data will help in determination of the extensions of the oil-zones and predicting future reservoir-behavior. This is an ongoing process accompanying the production activity throughout the reservoir age.

In studying the reservoir production energy, a special measurement yardstick is used. This is the (productivity index, PI), defined to be the number of barrels produced per day, corresponding to one pressure-unit of change. Naturally, the PI parameter is dependent upon the reservoir properties and conditions like permeability, gas saturation, production rate, and production lapse time. The production index normally falls with continued production, because of the fall in the reservoir pressure and increase of the oil viscosity resulting from loss of the dissolved gas.

When the reservoir is depleted at a too-high rate, (depletion rate higher than the rate of oil flow through the reservoir rocks), water and/or the gas found near the well, are sucked in and moved to the well itself, with result of production of these fluids in place of oil. This phenomenon (known as coning), causes quantities of oil to be blocked-in and hence, not possible to be produced. In order to avoid losing some of the oil reservoir and to secure maximum oil-recovery ratio, an optimum oil depletion rate must be applied.

7.6 Water and Gas Separation

During production, oil moves towards the lower-pressure zones of the reservoir which are near the producing oil-well. As oil is depleted the other two fluids (water and gas), in the reservoir rocks, move to the pores of the reservoir rocks, replacing the extracted oil in those pores. The rate of movement of the water and the gas depends largely on the reservoir dominating conditions. This, in turn, will have a direct effect on the flow rate of the oil and on the well productivity.

In general, the crude oil coming out of the producing well contains water, sediment, and dissolved gas. The percentages of these materials vary greatly from reservoir to reservoir. Water is usually present in greater quantities in old fields and in fields which have been subjected to water injection.

Before transporting the produced crude oil to export ports or to refineries, the associated water, gas and the sediment impurities are usually separated from oil. This is done by passing the raw crude oil into especially designed separation systems (the separator).

(i) Water Separation

The crude oil is normally expelled from the well head under high pressure (may exceed 1000 atm.). The water, associated with the produced oil, is normally characterized by its high salinity, and needs to be removed from oil before oil reaches export terminals or refineries. If not removed, the saline water can incur harms of the following types:

- Corrosion of the transporting pipes, tanks or any facility that water passes through.
- Precipitation of the salts on the inner surfaces of the heating-tube system.

- From economic point of view, leaving the water un-separated from oil shall incur extra burden on the transporting operation and storage facilities, as the water shall be eventually be discarded.

Usually the salty water is removed by special separating units usually referred to as (desalting units). Removing of water may be effected by gravity. That is passing the raw crude oil through a tank in which the water settles at the bottom where it is drained out.

During production, both of the water and oil move up the well in two phases; free water, and in very small water droplets dispersed in the oil liquid forming a mixture called (emulsion). Several ways lend themselves as means for the emulsion treatment. Some methods are based on heating of the emulsion, by chemical treatment, or by use of the electrical field. The separation process of the emulsion can be done by use of the hydro-cyclone mechanical principles which bring about separation by centrifugal-force effect.

(ii) Gas Separation

That part of the hydrocarbon accumulation which is in gaseous state (the natural gas) can be found in gas traps by its own, unassociated with liquid hydrocarbon (the oil), or as gas dissolved in oil, ready to be separated when the confining pressure is appropriately reduced. Raw natural gas produced from crude oil is normally referred to as the (associated gas).

A closely related to dry gas-oil matter is an intermediary phase of hydrocarbon, which is the gas-condensate or the wet gas, as it is sometimes called. Raw natural gas may be obtained from the associated gas (from crude-oil wells), from non-associated gas (from dry-gas wells), or from gas condensates (from gas-condensate wells). The natural gas condensates is a low-density hydrocarbon liquid which is originally a gaseous hydrocarbon present as a gas-component dissolved in the liquid oil, then condenses when temperature is reduced to a value below the hydrocarbon dew-point temperature of the hydrocarbon gas.

When the crude oil reaches the well head, it is relieved from the high reservoir pressure into the much lower atmospheric pressure. Consequently, the dissolved gas component gets separated from the liquid oil. At this stage the freed gas is either getting rid of by flaring or transporting it (after being subjected to certain processing operations) to marketing centers where it is used as thermal energy sources for domestic and industrial applications. Another important use of the produced gas is re-injecting it back into the oil reservoir to increase the reservoir production energy.

Gas separation is usually done by a special separation vessel (the gas-separator). An oil-gas separator is a pressure vessel installed near the well head for the purpose of separating the gas from the produced oil.

There are, in application, many types of oil-gas separators. Based on their mechanical configuration, separators can be of horizontal, vertical, or spherical separators. Based on their functions, they can be divided into two-phase (gas-oil separators) or three-phase (oil-gas-water separators). In general terms, the separator is a mechanical system designed to separate the hydrocarbon fluid, produced from an oil well, into its constituent components of gas, oil, and water. According to type of function, a separator is called deliquilizer for water removal and degasser (or degassing station) for gas removal.

7.7 Primary and Secondary Productions

At the start of oil production from a newly discovered oil field, oil rises through the production oil wells by the reservoir own pressure. This unaided production process is called (primary production). As production and oil depletion continue, reservoir energy (supported by pressure and temperature conditions) decreases with time until a state is reached whereby the reservoir energy becomes too low to push the oil through the well-pipes. At this state, oil production stops, although more oil is still there in the pores of the reservoir rocks.

The time when production stops, marks the end of the primary production stage which takes place by the reservoir naturally-provided energy. Duration of this production phase of production is dependent on the reservoir own conditions as well as on the applied production parameters. Thus, in order to produce the remaining oil, an additional, artificially-created energy is provided. With the additional energy-boosting measures, oil production is resumed. This, artificially-aided production is called (secondary or tertiary recovery).

The main factor that sustains the reservoir energy is the pressure imposed by the hydrocarbon gases (in-oil dissolved or from-oil liberated gases) in addition to formation water. According to the principal drive agent, it is decided to boost the reservoir energy by either injecting water into the formation-water zone or by injecting gas in the gas-cap. Decision on the type of method to enhance oil recovery is based on whether the principal drive agent is water drive or gas drive.

(i) The Water Injection Method

This method is the more commonly applied way used in oil secondary recovery. Based on studies made on the reservoir, water is injected in a number of wells of carefully measured depths, and drilled in defined locations. The injected water moves in the direction of the oil zone displacing and pushing the oil in the direction of the production wells through which it rises to the well heads.

(ii) The Gas Injection Method

In cases where there are difficulties in the water injection process, gas is used instead of water. Gas injection can help boosting oil production in two ways. One way is effected by re-pressurizing of the gas cap, and also by its effect in decreasing the oil viscosity which occurs as a result of dissolution of part of the injected gas in the liquid oil.

(iii) The Air or Steam Injection Method

Instead of water or hydrocarbon gas, the injected matter can be burned-air, high-temperature steam, or heated water. In such cases the high temperature causes increase of pressure due to thermal expansion of the oil body. Consequently oil is moved to the lower-pressure zones near the drilled production wells.

(iv) The Permeability-Increasing Methods

To increase oil production, there is a group of methods depending on improving the permeability of the reservoir rocks to facilitate oil passage to production wells. One of such procedures is done by injecting high-pressure fluids which act on fracturing rock texture. Another way of improving the permeability is done by applying chemical means, as injecting acids which help in dissolving parts of the reservoir rocks, especially when they are of carbonate-type. Both of these methods (rock fracturing and acidizing) act as permeability-improving agents and hence enhancing oil recovery.

Procedures of artificial energy-boosting of oil reservoirs for the purpose of improving oil-productivity are classified into secondary and tertiary methods. Those methods which are depending on injecting water, gas, or air at normal temperature levels are usually referred to as (the secondary recovery) methods. The other type of procedure, based on using thermal burning or using chemical reactions (as acidizing), are called (tertiary recovery methods). Because of the basically high-cost operations, tertiary recovery methods are much less common in application than the secondary recovery methods.

(v) Use of Pumping Methods

In certain cases, when the reservoir energy reaches such a low level that the oil, collected at the bottom of the hole, cannot be pushed up to surface. In this case oil lifting is effected by us of mechanical or electrical pumping methods. One common mechanical pumping system is the so-called (Nodding Donkey Pump), as it is schematically shown in (Fig. 7.1).

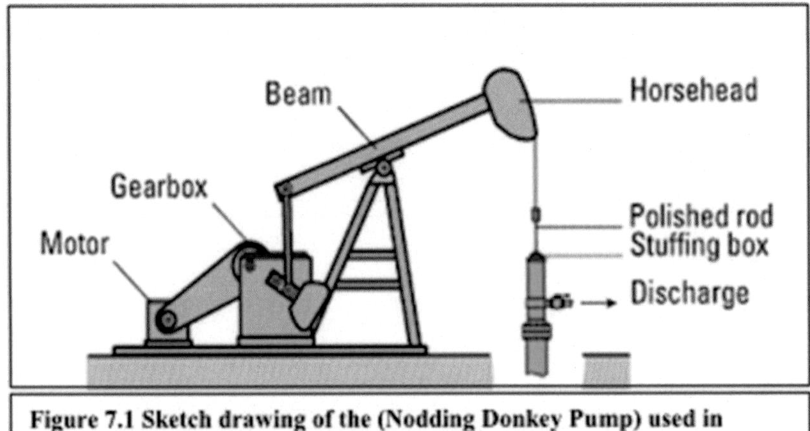

Figure 7.1 Sketch drawing of the (Nodding Donkey Pump) used in pumping oil from low-energy oil-reservoirs

The mechanical framework of the Nodding Donkey pump consists of a metal beam (with an end-mass termed the "Horse-head") connected by a rod, to a pump placed at the bottom of the hole. By running a motor, connected to the beam (by a lever system), an up-and-down motion is conveyed to the down-hole pump assembly. The so-created pumping action oil is forced to rise through the production pipes.

Another way of producing oil from low-energy oil reservoirs is the use of submersible electrically driven pumps. A specially designed pump assembly is lowered inside the oil well. The submersible pump-assembly consists of two main parts; the pump and an electric motor that provides the power needed for pump operation.

7.8 World Oil and Gas Production

Based on the annual statistical bulletin published in 2017 by OPEC, oil and gas world production statistics (for 2016) are here presented. Oil and natural gas production rates are given, in descending order, for the world top 50 countries.

(i) The World Oil Production

Oil production daily rate for the world top 50 countries (in 2016), is given in the following table.

No	country	million barrel/day
1	Saudi Arabia	10.460
2	Russia	10.292
3	Venezuela	2.373
4	United States	8.875
5	Iraq	4.648

No	country	million barrel/day
6	China	3.982
7	Iran	3.651
8	United Arab Emirates	3.088
9	Kuwait	2.954
10	Brazil	2.510
11	Mexico	2.154
12	Angola	1.722
13	Norway	1.616
14	Nigeria	1.427
15	Kazakhstan	1.295
16	Canada	1.186
17	Algeria	1.146
18	United Kingdom	0.915
19	Oman	0.909
20	Colombia	0.883
21	Azerbaijan	0.770
22	Indonesia	0.740
23	India	0.721
24	Malaysia	0.670
25	Qatar	0.652
26	Ecuador	0.549
27	Egypt	0.544
28	Argentina	0.512
29	Libya	0.390
30	Congo	0.306
31	Australia	0.289
32	Sudan & South Sudan	0.265
33	Equatorial Guinea	0.236
34	Gabon	0.220
35	Bahrain	0.205
36	Turkmenistan	0.190
37	Denmark	0.140
38	Brunei	0.109
39	Romania	0.076
40	Trinidad & Tobago	0.072
41	Italy	0.071
42	Turkey	0.050
43	Germany	0.046
44	Peru	0.040

No	country	million barrel/day
45	New Zealand	0.035
46	Ukraine	0.032
47	Yemen	0.024
48	Netherlands	0.018
49	Syria	0.017
50	France	0.017

Table of the oil daily production (in 2016) for the top 50 countries ((OPEC Annual Statistical Bulletin-2017).

(ii) The World Gas Production

Gas production annual rate for the world top 50 countries (in 2016), is given in the following table.

No	country	billion cubic meter
1	United States	751.063
2	Russia	642.242
3	Iran	226.905
4	Qatar	182.830
5	Canada	174.051
6	China	136.628
7	Norway	120.193
8	Saudi Arabia	110.860
9	Algeria	93.152
10	Turkmenistan	81.765
11	Indonesia	74.026
12	Malaysia	64.428
13	United Arab Emirates	61.084
14	Uzbekistan	57.700
15	Australia	56.293
16	Netherlands	50.543
17	Trinidad & Tobago	43.374
18	United Kingdom	43.022
19	Nigeria	42.562
20	Pakistan	42.209
21	Egypt	42.102
22	Mexico	41.227
23	Thailand	38.929
24	Argentina	36.546

No	country	billion cubic meter
25	Oman	32.779
26	India	31.139
27	Venezuela	27.718
28	Bangladesh	25.512
29	Bolivia	23.501
30	Bahrain	22.351
31	Kazakhstan	22.004
32	Brazil	20.619
33	Ukraine	19.271
34	Azerbaijan	18.773
35	Myanmar	18.529
36	Kuwait	17.291
37	Libya	15.571
38	Peru	14.454
39	Colombia	12.935
40	Brunei	11.132
41	Iraq	10.416
42	Romania	9.890
43	Vietnam	9.298
44	Germany	7.606
45	Equatorial Guinea	6.212
46	Mozambique	5.826
47	Poland	5.794
48	Italy	5.783
49	New Zealand	5.063
50	Denmark	4.505

Table of the gas annual production (year 2016) for the world top 50 countries as reported by OPEC Annual Statistical Bulletin-2017.

According to this reference (OPEC 2017-bulletin), the world total oil production-rate (in 2016) is 75.477 million barrel/day and the corresponding figure for the world marketed gas production is 3674.471 billion cubic meter (= 3.675 tcm).

Chapter-8

8. OIL TRANSPORT AND STORAGE

Oil, flowing out from production wells is, after passing through gas separators, transported through a system of pipelines to storage tanks in the oil field. From these tanks oil is then pumped through pipes, of larger diameter, to a tank farm built as a part of an oil-refinery or to a costal tank farm where it is stored and be made ready for export. In short, oil needs to be transported as crude matter coming out of the well mouth, and as refined products outputted from refineries.

There are many ways to transport crude oil, or oil derivatives. On-land transportation can be done by trucks, by special on-rail tanker-cars, or by on-land pipes. Sea transportation, The principal method used in marine transportation is use of oil tankers of various sizes. Choice of the transportation means is governed by the nature of the transported matter as well as the total transportation environments.

8.1 Transports by Trucks

Trucks equipped with oil storage tanks are normally used in moving quantities of oil (crude or derivatives) over roads and highways to certain planned destinies. In general, truck transportation is used when oil quantities are relatively small and over relatively short distances (Fig. 8.1).

Figure 8.1 Tanker truck for oil transportation.

Usually, the method is used in distributing oil derivatives (as gasoline, kerosene, diesel fuel) to petrol stations. In certain cases crude oil is transported by tanker trucks.

8.2 Transports by On-Rail Tankers

Specially equipped cargo trains are used in transporting oil. Such trains are made up of tanker carriages (rail cars) similar to the truck tankers but usually of larger sizes (Fig. 8.2).

Figure 8.2 On-Rail tanker carriages for oil transportation.

Compared with the truck method, rail transportation is far more effective way in transporting huge amounts of oil across more extensive land areas. It can be used in transporting great amounts of crude oil (or oil derivatives) long distances in open lands.

8.3 Transports by Pipelines

Pipeline transportation is considered to be the most cost-effective means of oil. In an oil field, production wells are normally connected by a network of pipelines. In this way the oil, produced from the wells, is moved to degassing stations, and then pumped to a storage tank-farm built near a refinery or near a sea coast. Usually, an oil pipeline is built to move oil hundreds or thousands of kilometers from oil field to a tank-farm (Fig. 8.3).

Pipelines are usually equipped with special booster pumps which are providing the necessary energy for the oil to keep moving through the entire pipe length.

Figure 8.3 Giant Pipeline for oil transportation.

As it is the case with the other methods, pipelines can be used in transporting oil derivatives, although they are mainly used for transporting crude oil. Different types of crude oil, or of oil derivatives, can be moved through the same line simultaneously with negligible intermixing or contamination. This is done by using special plastic balls as separators between the different liquids.

8.4 Transports by Oil Tankers

An oil tanker is a large ship especially designed to carry oil (mainly crude oil) across oceans. The hull of the oil tanker is divided into a number of oil-tight compartments into which oil is pumped when the tanker is being loaded. For oil loading, on-shore pumps are used whereas on-ship pumps are used for oil discharge (Fig. 8.4).

Figure 8.4 Large ship (oil tanker) for oil transportation.

Oil tankers vary in sizes, with large vessels of (100 000 – 200 000) tons are currently used for global-scale transportation of crude oil. At present, super-large tankers (in excess of 300 000 tons) are in service (Smil, 2008). Other smaller-size marine vessels (oil tankers and barges) are generally used to transport oil derivatives over relatively short distances in seas, through rivers, and canals.

8.5 Oil Storage Tank-Farm

A tank farm is an industrial installation made up of a group of metal tanks for oil storage. A tank farm is usually built near refineries, at pipeline terminals, and at locations where oil-carrying tankers or ships can take on, or discharge, their loaded oil. The storage tanks in these farms are so constructed that they can be filled in and emptied efficiently and with no harm to environment.

Tank farms are of various types and sizes. The usual form is the vertical, cylindrical tanks which may be open-top, fixed-roof, or floating-roof tanks (Fig. 8.5).

Figure 8.5 Tank-farm for oil storage.

Photo courtesy Crowley Petroleum Distribution

A tank content of oil can be measured manually or with automatic tank gauges. Tank measurements provide information on oil quantity and level height in each tank. This is important for control of the total oil stock and to take the appropriate measures for preventing overfilling and possibility of oil spill.

REFERENCES

Al-Sadi, H. N., 1980, Seismic Exploration, Technique and Processing. 215p, Birkhauser Verlag, Basel, Switzerland.

Alsadi, H. N., 2017, Seismic Hydrocarbon Exploration, 331 p., Springer International Publishing AG, Switzerland.

Assayyab, A. & Abdulhameed, M, 1979, Petroleum Geology (in Arabic language), published by Baghdad University, Baghdad, Iraq.

Banks, R. E. & King, P. J., (in book edited by Hobson, G. D.,), 1986, Modern Petroleum Technology, Ed.5, J. W. Wiley & Sons, by the Institute of Petroleum, London.

Chapman, R., E., 1976, Petroleum Geology, Elsevier, Amsterdam.

Dobrin, M. B., and Savit, C. H., 1988, Introduction to Geophysical Prospecting, McGraw Hill Book Company, 867p, New York.

Dobrin, M. B., 1960, Introduction to Geophysical Prospecting, McGraw Hill Book Company, 446p, New York.

Eramo, M., 2015, Inaugural Ethylene Forum, Technip Engineering Company.

Fairhead, J. D. (1988) Mesozoic plate tectonic reconstruction of the central south Atlantic Ocean: The role of the west and central African rift system. Tectonophysics, 155, pp. 181-195.

Goddard, F. W., & Hutton, K., 1955, A School Chemistry for Today, Longmans, London.

Gutenberg, 1959, Physics of the Earth's Interior, Academic Press

Halbouty, M. T. et al, 1970, World's Giant Oil and Gas Fields. AAPG Mem14, p 502-556.

Hobson, G. D.,1986, The occurrence and origin of oil and gas, in Modern Petroleum Technology, part-1, 5th ed., John Wiley.

Holmes, A. 1965, Principles of Physical Geology, Nelson, 1288p.

Hyne, N. J., 2012, Petroleum Geology, Exploration, Drilling, & Production (3rd. ed.), Pennwell Corporation., Tulsa, Oklahoma, USA, 698 p.

Kearey, P., Brooks, M., and Hill, I., 2002, An Introduction to Geophysical Exploration, Blackwell Science Ltd., Oxford, 262p.

Prentice Hall Publication, 1995, Exploring Physical Science, Prentice Hall Publishing Company, Englewood Cliffs, New Jersey, USA.

Schull, T. J. (1988) Rift basins of interior Sudan: Petroleum exploration and discovery. American Association of Petroleum Geologists Bulletin (AAPG), 72, pp. 1128-1142, Tulsa.

Selley, R. C., 1983, Petroleum Geology for Geophysicists and Engineers, IHRDC, Boston, 88p.

Sheriff, R. E., and Geldart, L. P., 1995, Exploration Seismology, Cambridge University Press, 592 p., UK.

Smil, V., 2008, Oil, A Beginner's Guide, Oneworld Publications, Oxford, 592 England.

Stoneley, R., 1995, Introduction to Petroleum Exploration for Non-Geologists, Oxford University Press Inc., 119 p.New York, USA.

Stringer, R. and Johnston, P., 2002, Chlorine and the Environment, Kluwer academic Publishers, 437 p.,The Netherlands..

Wymer, N., 1964, Behind the Scenes in an Oilfield, Dent & Sons Ltd., London, UK.

Books Recommended for further reading

Out of the vast number of books written in the field of Petroleum Science, I chose the following books which are dealing with the Petroleum principal scientific branches.

The selected books are listed here-below:

1. **Petroleum Chemistry**

Speight, J. G., 1998, Petroleum Chemistry and Refining, CRC Press, USA, 350 p.

Speight, J. G., 2014, The Chemistry and Technology of Petroleum (5$^{th.}$. Ed.),, Talor & Francis Group, USA, 953 p.

Matar, S. and Lewis, F. H., 2001, Chemistry of Petrochemical Processes (2nd. Ed.), Gulph Professional Publishing Co. 356 p.

Hunt, J. M., 1996, Petroleum Geochemistry and Geology, (2nd. Ed.), Freeman Publishing Company, San Francisco, USA.

2. **Petroleum Geology**

Selley, R. C. and Sonnenberg, S. A., 2015, Elements of Petroleum Geology (3rd. ed.), Academic Press (Imprint of Elsevier), 514 p.

Levorsen, A. I., 1967, Geology of Petroleum, Freeman Co., San Francisco, USA.

Chapman, R., E., 1976, Petroleum Geology, Elsevier, Amsterdam.

Selley, R. C., 1998, Elements of Petroleum Geology (2nd. Ed.), Academic Press Limited, San Diego, USA, 470 p.

Hyne, N. J., 2012, Petroleum Geology, Exploration, Drilling, & Production (3rd. ed.), Pennwell Corporation., Tulsa, Oklahoma, USA, 698 p.

3. **Petroleum and Petrochemical Engineering**

Archer, J. S. and Wall, C. G., 1986 (reprint 1994), Petroleum Engineering Principles and Practice, graham and Trotman ltd., London.

Chaudhuri, U. R., 2011, Fundamentals of Petroleum and Petrochemical Engineering, CRC Press, Taylor & Francis Group, 411 p.

Burdick, D. L. and Leffler, W. L., 2001, Petrochemicals in Nontechnical Language, Penwell Publishing Company, 450 p., Tulsa, USA.

Banks, R. E. & King, P. J., (in book edited by Hobson, G. D.,), 1986, Modern Petroleum Technology, Ed.5, J. W. Wiley & Sons, by the Institute of Petroleum, London.

4. **Petroleum Exploration**

Alsadi, H. N., 2017, Seismic Hydrocarbon Exploration, 331 p., Springer International Publishing AG, Switzerland.

Sheriff, R. E., and Geldart, L. P., 1995, Exploration Seismology, Cambridge University Press, 592 p., UK.

Stoneley, R., 1995, Introduction to Petroleum Exploration for Non-Geologists, Oxford University Press Inc., New York, USA.

5. **Petroleum-Well Drilling**

Bommer, P. M., 2008, A Primer of Oil well Drilling, University of Texas Press, 241 p.

Austin, E. H., 1983, Drilling Engineering Handbook, IHRDC, Boston, 301 p.

6. **Petroleum Origin**

Altovoskii, M. E., 1961, Origin of Oil and Oil Deposits, Springer Science+Business,107 p., NY,USA.

Tissot, B. P. and Welte, D. H., 1978, Petroleum Formation and Occurrence, Springer-Verlag, Berlin,

Hobson, G. D.,1986, The occurrence and origin of oil and gas, in Modern Petroleum Technology, part-1, 5th ed., John Wiley.

7. **Petroleum Production**

Devold H., 2009, Oil and Gas Production Handbook (2nd. Ed.), ABB-Oil & Gas, Oslo, Norway, 113 p.

Guo, B., Liu, X., and Tan, X, 2017, Petroleum Production Engineering (2nd. Ed.), Elsevier Inc. (Imprint Gulph Professional Publishing Co.), 780 p.

8. **Petroleum Refining**

Leffler, W. L., 2008, Petroleum Refining in Nontechnical Language (4th. Ed.), Penwell Corporation, 780 p., Tulsa, USA.

9. **Unconventional Hydrocarbon Resources**

Ma, Z. Y., and Stephen, A. H., 2016, Unconventional Oil and Gas Resources Handbook, Gulph Professional Publishing Co. (Imprint of Elsevier), 550 p.

Usman, A. and Meehan N, 2016, Unconventional Oil and Gas Resources, Exploitation and Development, CRC Press, Taylor & Francis Group, 860 p.

Yen, T. F., and Chilingarian, G. V., 1976, Oil Shale, Elsevier, 291 p.

Chilingarian, G. V. and Yen, T. F., 1978, Bitumen Asphalts and Tar Sands, Elsevier, Amsterdam.

Gayer, R. and Harris, I., 1996, Coalbed Methane and Coal Geology, Spec. Publ. Geol. Soc. London.

APPENDICES

A-1 Counting Prefixes

measurements unit prefixes	symbol	Units for length measurements
tera	T	1 terameter (Tm) = 10^{12} m
Giga	G	1 gigameter (Gm) = 10^{9} m
mega	M	1 megameter (Mm) = 10^{6} m
kilo	k	1 kilometer (km) = 10^{3} m
hecto	h	1 hektometer (hm) = 10^{2} m
deca	da	1 decameter (dam) = 10^{1} m
		1 meter (m) = 10^{0} m
deci	d	1 decimeter (dm) = 10^{-1} m
centi	cm	1 centimeter (cm) = 10^{-2} m
milli	mm	1 millimeter (mm) = 10^{-3} m
micro	μ	1 micrometer (μm) = 10^{-6} m
nano	n	1 nanometer (nm) = 10^{-9} m
pico	P	1 picometer (pm) = 10^{-12} m

It is to be noted that we have the following alternative names:
megameter = million meter, gigameter = billion meter, and terameter = trillion meter.

A-2 Conversions for Units of Length

unit	centimeter	meter	inch	foot	yard
1 centimeter =	1.0	0.01	0.3937	0.0328	0.0109
1 meter =	100.0	1.0	39.37	3.281	1.0937
1 inch =	2.54	0.0254	1.0	0.0833	0.0278
1 foot =	30.48	0.3048	12.0	1.0	0.3333
1 yard =	91.44	0.914	36.0	3.0	1.0

Also we have, one international nautical mile = 6076.12 feet = 1.852 kilometer (km), where one mile is equal to 1760 yards (=5280 feet =1609.34 meter).

A-3 Conversions for Units of Area

unit	square centimeter	square meter	square inch	square feet	square yard
1 square centimeter =	1.0	0.0001	0.155	0.001076	0.00012
1 square meter =	10000.0	1.0	1550.0	10.76	1.196
1 square inch =	6.4516	0.000645	1.0	0.00694	0.000772
1 square feet =	929.030	0.09290	144.0	1.0	0.1111
1 square yard =	8364.3	0.8361	1296.0	9.0	1.0

We have one square kilometer (= 0.386 square mile). Further, we have:

1 acre = 1/640 square mile = 43560.0 square foot = 4046.855 square meter. Further:

1 hectare = 10000 square meter 2.471 acres.

A-4 Conversions for Units of Volume

unit	cubic cm. (cm^3)	cubic m. (m^3)	liter	cubic inch (in^3)	cubic foot (ft^3)	gallon (US)
1 cubic cm =	1.0	$1.0x10^{-6}$	0.001	0.0610	$3.53x10^{-5}$	$2.64x10^{-4}$
1 cubic m =	$1.0x10^{6}$	1.0	1000	61.023	35.31	264.2

unit	cubic cm. (cm^3)	cubic m. (m^3)	liter	cubic inch (in^3)	cubic foot (ft^3)	gallon (US)
1 liter =	1000	0.001	1.0	61.024	0.0353	0.2642
1 cubic inch =	16.38	$1.638x10^{-5}$	0.0164	1.0	$5.764x10^{-4}$	0.0043
1 cubic foot =	$2.832x10^4$	0.02832	28.32	1728	1.0	7.482
1 gallon (US) =	3785	$3.785x10^{-3}$	3.785	230.97	0.1337	1.0

Regarding the gallon unit, there are more than one definition. The commonly known definitions are two; the imperial (British) gallon is equal to 4.546 liter and the US gallon which is equal to 3.785 liter. This means that the imperial gallon = 1.2 US gallon.

The other volume unit, commonly used in petroleum measurements, is the (barrel), defined to be equal to 42 US gallon = 35 UK (imperial) gallon, approximately equal to 159 liter (5.6 cubic foot). To convert volume to mass the density factor is needed, For example, one ton of Brent crude oil (38.06 API, Sp. Gr.=0.835) will be 7.532 barrel.

A-5 Conversions for Units of Mass

units	gram	kilogram	once	pound	metric ton
1 gram =	1.0	10^{-3}	0.03527	0.0022	0.000001
1 kilogram =	1000	1.0	35.273	2.205	0.001
1 once =	28.35	0.02835	1.0	0.0625	$2.835x10^{-5}$
1 pound =	453.6	0.4536	16.0	1.0	$4.536x^{-4}$
1 metric ton =	10^6	1000	35273.6	2204.6	1.0

In weight measurements, very often one comes across the two units (long and short tons). The long ton (an imperial measurements system) is defined as a mass of exactly 2240 pound. In the traditional British measurements system, it is equal to 20 hundred-weight units (abbreviated Cwt), where each unit is 8 stones and one stone is equal to 14 pound. The short ton is defined to be equal to 2000 pound. Clearly, the long ton is equal to 1.12 short ton.

A-6 Petroleum Specific Gravity – API Degrees

API degrees	Specific Gravity	API degrees	Specific Gravity
10	1.000000	31	0.870769
11	0.992982	32	0.865443
12	0.986063	33	0.860182
13	0.979239	34	0.854985
14	0.972509	35	0.849850
15	0.965870	36	0.844776
16	0.959322	37	0.839763
17	0.952862	38	0.834808
18	0.946488	39	0.829912
19	0.940199	40	0.825073
20	0.933993	41	0.820290
21	0.927869	42	0.815562
22	0.921824	43	0.810888
23	0.915858	44	0.806268
24	0.909968	45	0.801700
25	0.904153	46	0.797183
26	0.898413	47	0.792717
27	0.892744	48	0.788301
28	0.887147	49	0.783934
29	0.881620	50	0.779614
30	0.876161	51	0.775342

Conversion of an API value to the corresponding specific gravity is done through application of the formula (API = 141.5/Sp. Gr. – 131.5).

A-7 Conversions for Units of Pressure

units	gm/cm²	psi	psf	pascal	bar	atm	cm water	cm mercury
1 gm/cm² =	1.0s	0.0142	2.048	97.9	0.00098	0.00097	1.0	0.0735
1 psi =	70.42	1.0	144.0	6894.7	0.069	0.068	70.56	5.184
1 psf =	0.488	144.0	1.0	47.9	0.00048	0.00047	0.488	0.036
1 pascal =	0.0102	0.000145	0.021	1.0	0.00001	0.00001	0.010	0.00075
1 bar =	1021.4	14.5	2088.6	100 000	1.0	0.99	1023.1	75.164
1 atm =	1038.9	14.7	2116.8	101317.1	1.0138	1.0	1037.2	74.412

Pressure is defined to be the force which is perpendicularly applied to unit area of a surface. Thus units can be gram weight per square centimeter (gm/cm^2), kilogram weight per square meter (kg/m^2), pound weight per square inch (psi), or pound weight per square foot (lb/ft^2). Other derived units, in common use, are: Pascal (Pa), atmosphere (atm), or barometric units as, centimeter (or meter) of water- and inch mercury-column heights.

The Pascal unit is an another commonly used pressure unit, defined to be equal to one Newton per square meter (N/m^2) and the atmosphere unit (atm) is the pressure exerted by the natural air on Earth surface at sea level, which is about 14.7 psi. The bar unit of pressure measurements, is defined to be equal to 100 000 Pa (100 000 N/m^2).

The way to calculate the column height units (cm-water, cm-mercury, …) can be done from the relationship (pressure = density x column height), using the appropriate density units. The other pressure units conversions are found from the physical definition of the units.

A-8 Fahrenheit – Celsius Degrees Relationships

Fahrenheit degrees	**Celsius degrees**
104	40
95	35
86	30
77	25
68	20
59	15
50	10
41	5
32	0
23	-5
14	-10
5	-15
0	-17.8

For any other temperature values, the following formulae are used to convert Fahrenheit degrees (F^0) to Celsius degrees (C^0) and vice versa:

$$(F^0) = (9/5)\ C^0 + 32$$

$$(C^0) = (5/9)\ (F^0 - 32)$$

A-9 Abbreviations

This is a list of abbreviations, often found associated with petroleum academic and industrial activities.

AAPG
American Association of Petroleum Geologists

ADPC
Abu Dhabi Petroleum Company

API
American Petroleum Institute

Barrel
Unit of measuring oil quantities (= 159 liters = 35 imperial gallons = 42 US gallons)

B/D
Measurement unit for the rate of petroleum flow, in barrel per day (b/d), as in oil production

BOP
blowout preventer

CAJG
Conference of the Arabian Journal of Geosciences

CFP
Compagnie Francaise des Petrol

CGG
Compagnie Generale de Geophysique

CNLC
China national logging corporation

CNPC
China national petroleum corporation

CNPC
China national petroleum corporation

DHI
Direct hydrocarbon indicator (exploration tool based on data from seismic reflection surveying)

DST
Dril stem test (test to evaluate hydrocarbon-production capability of a drilled rock formation)

EAEG
European Association of Exploration

EOR
Enhanced oil recovery

E & P
Exploration and production

ESSO
It is an acronym for Eastern States standard Oil. It was used as a trading name for Standard Oil of New Jersey, and on 1. 11. 1972, it was changed to EXXON.

GOC
Gas-oil contact

GOR
Gas-oil ratio, cubic feet of gas per barrel of oil in a subsurface oil reservoir

GSA
Geological Society of America

GSI
Geophysical Service Incorporated

GWC
Gas-water contact

IFP
Institute Francais du Petrole (French institute for petroleum science research)

KB
Kelly bushing (part of the rotary drilling system of an oil-well drilling rig)

KOC
Kuwait Oil Company

LNG
Liquefied natural gas

LOM
Level of organic maturation (maturation of kerogen found in a source rock)

LPG
Liquefied petroleum gas

LUKOIL
Russian multinational oil and gas company

MBD
Million barrel per day (mb/d)

NIOC
National Iranian Oil Company

OAPEC
Organization of Arab Petroleum Exporting Countries

OPEC
Organization of Petroleum Exporting Countries

OWC
Oil-water contact

PEMEX
Petroleos Mexicanos, national oil company of Mexico

PERTAMINA
Pertambangan Minjak dan Gas Bumi Negara (the Indonesian state oil and natural gas company)

PETRONAS
Malaysian national oil company

SEG
Society of Exploration Geophysicists

SI
Systeme International (for measurement units)

SOECOR
Southern Oil Exploration Corporation, South African State Oil Cpmpany.

SPE
Society of Petroleum Engineers

SSC
Seismograph Service Corporation

STATOIL
Norwegian state oil Company

TD
total depth

TI
Texas Instruments

TOTAL
French multinational oil and gas company

VSP
Vertical Seismic Profiling

Index

A

B

C

P

R

S

T

U

V

W

Printed in the United States
By Bookmasters

This book is a concise presentation of the science of petroleum. It covers the basic elements of the Petroleum science through eight chapters. The first chapter contains basic definitions together with the chemical composition and physical properties of the petroleum substance. The second chapter is a summary of those geological concepts pertinent to the petroleum habitat rocks. In the following two chapters, petroleum generation, migration, and accumulation into petroleum reservoirs, are described. Petroleum exploration techniques, well drilling, production and storage, are dealt with in the last four chapters.

The book is designed to serve audiences from both the academic and industrial worlds. University students and staff members of oil-exploration companies will find this book very helpful in increasing their knowledge and in boasting their application effort's efficiency.

I will be grateful for readers who can let me know of any comment of criticism. Such contributions shall be used in improving future book updates. My email address is hamidalsadi@hotmail.com.

• • • • • • • • • • • • • • • •

The author, **Hamid Nassar Alsadi**, is a native of Hit, Alanbar province, Iraq. He obtained his BSc. in Geology, Physics, and mathematics from Reading University, England, MSc. in Geolphysics from Durham University, England, and PhD. in Seismology from Uppsala University, Sweden. Since 1972, he was mainly working in petroleum exploration using 2D and 3D seismic surveying. During this time, he was equally active in the academic domain, where he conducted teaching courses, thesis supervision for university postgraduate students, and published a number of books and papers dealing with petroleum geophysical exploration. After retirement from his original post in the Iraqi state oil companies, he worked as a consultant with oil exploration establishments at home and abroad, including giving numerous (> 45) training courses for oil-exploration personnel.

ISBN 978-1-5437-4814-7
90000
9 781543 748147

Scooter Challenges the Day

A True Story as told by Patrick Trotter

By Marielle D. Marne
Illustrated by Herb James